SD 387 R4 W89

Understanding
Forest Disturbance and
Spatial Pattern

Remote Sensing and GIS Approaches

Understanding Forest Disturbance and Spatial Pattern

Remote Sensing and GIS Approaches

Edited by
Michael A. Wulder
Steven E. Franklin

Taylor & Francis
Taylor & Francis Group
Boca Raton   London   New York

CRC is an imprint of the Taylor & Francis Group,
an informa business

CRC Press
Taylor & Francis Group
6000 Broken Sound Parkway NW, Suite 300
Boca Raton, FL 33487-2742

© 2007 by Taylor & Francis Group, LLC
CRC Press is an imprint of Taylor & Francis Group, an Informa business

International Standard Book Number-10: 0-8493-3425-X (Hardcover)
International Standard Book Number-13: 978-0-8493-3425-2 (Hardcover)

Library of Congress Cataloging-in-Publication Data

Wulder, Michael A.
    Understanding forest disturbance and spatial pattern : remote sensing and GIS approaches / author / editors, Michael Wulder and Steven E. Franklin.
        p. cm.
    Includes bibliographical references and index.
    ISBN 0-8493-3425-X (alk. paper)
    1. Forests and forestry--Remote sensing. 2. Geographic information systems. I. Franklin, Steven E. II. Title.

SD387.R4W85 2006
634.9--dc22   2006007999

Visit the Taylor & Francis Web site at
http://www.taylorandfrancis.com

and the CRC Press Web site at
http://www.crcpress.com

To Yoka and John

M.A.W.

For Barb

S.E.F.

Preface

This book was conceived as a contribution to the increasingly urgent need in the scientific and resource management communities to develop greater understanding of forest disturbance as a means to aid in the resolution of complex forest management planning issues worldwide. It is our view that, in the past decade, the capability to use remotely sensed data for the generation of forest disturbance products is increasingly well understood and, consequently, more widely available. The "how-to" questions that have preoccupied geospatial analysts and practicing resource management professionals are now less critical. Rather, clarification is sought on the wider ecological meaning of the spatial patterns associated with disturbance and what can and should be done with the copious and diverse information that is generated by remote sensing and geographical information system (GIS) approaches. In addition, questions are emerging regarding how forest practices should be changed (if at all) to accommodate the new perspectives generated by geospatial technologies. For example, the use of landscape metrics to characterize landscape pattern from remotely sensed map products enables unprecedented opportunities for improved forest management and sustainable stewardship. Landscape metrics provide a synoptic and systematic means to understand the implications of disturbance processes, including issues such as altered habitat and forest fragmentation. However, the appropriate application and insightful interpretation of landscape metrics are only in their infancy, as are the emerging disciplines of landscape ecology and conservation biology, both of which owe a portion of their growth and potential to developments in the fields of remote sensing and GIS.

We perceived an opportunity to present in this book a sequence of topics that would take the reader from a general biological or landscape ecological context of forest disturbance, to remote sensing and GIS technological approaches, through to pattern description and analysis, with compelling applied examples of integration and synthesis. The chapters for this volume were invited, peer reviewed, revised, and edited; the authors and reviewers adhered to the strictest standards and highest quality criteria in this process. The issues discussed here address both natural and human-caused forest change and include factors such as biological components, monitoring approaches, scale, and pattern analysis. In this book, our goal was to consider forest disturbance and spatial pattern from an ecological point of view within the context of structure, function, pattern, and change. Remotely sensed and GIS data are now the data sources of choice for those whose responsibility it is to capture, document, and understand landscape pattern and forest disturbance. A discussion of the concepts of pattern characterization, which is an area of research and application we expect will continue to grow in importance and significance to resource managers, highlights the challenges in this emerging area of research, and although significant progress has been made, clearly much remains to be done. We conclude this book with a final chapter in which we provide a summary and description of the thematic issues related to detection and mapping of forest disturbances with remotely sensed and GIS data. Over the course of the book, we attempt to illustrate how the elements presented from ecological underpinnings, data considerations, change detection method, and pattern analysis, combine as a problem-solving, information-generating approach. It is our hope that the materials presented will stimulate discussion and provide guidance for those who are interested, or faced with similar challenges, in capturing and characterizing forest disturbance and pattern.

ACKNOWLEDGMENTS

We acknowledge the important financial and administrative support from the Canadian Forest Service, the University of Saskatchewan, and the Natural Sciences and Engineering Research Council of Canada, without which this book would not have been possible. Sincere thanks are likewise extended to the many authors and reviewers for their enthusiastic participation in this project. Any contribution that this volume will make in advancing the understanding of spatial pattern and forest disturbance through remote sensing and GIS approaches is largely a result of their effort and commitment. A special thank you is offered to Jill Jurgensen and the staff and editors at Taylor & Francis for their superb assistance in bringing the source manuscript to fruition in this book publication. David Seemann and Joanne White of the Canadian Forest Service are thanked for support and assistance with the production of the book manuscript. To our many collaborators and colleagues, we are appreciative and indebted for your insights and constructive engagement on so many aspects of research, development, and communications. Finally, and especially, to our families, we express our sincere thanks for love, understanding, and continued support.

Michael A. Wulder
Victoria, British Columbia

Steven E. Franklin
Saskatoon, Saskatchewan

About the Editors

Michael A. Wulder received his B.Sc. (1995) degree from the University of Calgary and his M.E.S. (1996) and Ph.D. (1998) degrees from the Faculty of Environmental Studies at the University of Waterloo. On graduation, Dr. Wulder joined the Canadian Forest Service, Pacific Forestry Centre, in Victoria, British Columbia, as a Research Scientist. He has served on the executive committee of the Canadian Remote Sensing Society since 2003, was co-chair of the 22nd Canadian Symposium on Remote Sensing in 2000, and co-edited the ensuing special issue of the *Canadian Journal of Remote Sensing* (December 2001). He has also co-chaired international workshops on light detection and ranging (LIDAR) remote sensing of forests in Canada and Australia (which also resulted in a special issue of the *Canadian Journal of Remote Sensing* in October 2003). In 2000, he assumed the lead role in the Land Cover component of the Earth Observation for Sustainable Development of forests program, funded by the Canadian Space Agency and the Canadian Forest Service. A land cover map of the forested area of Canada is nearing completion based on this effort. Dr. Wulder's research interests focus on the application of remote sensing and GIS technologies to address issues of forest structure and function. Dr. Wulder's major research publications include the book *Remote Sensing of Forest Environments: Concepts and Case Studies* (2003, KAP) and a forthcoming book entitled *Monitoring of Large Area Forest Cover and Dynamics* (2007, Taylor and Francis). He has produced more than 60 refereed articles in leading journals from the fields of remote sensing, forestry, environmental science, and geographical information systems. Dr. Wulder is an adjunct professor in the Department of Geography at the University of Victoria and the Department of Forest Resources Management of the University of British Columbia.

Steven E. Franklin received his B.E.S. (1980), M.A. (1982), and Ph.D. (1985) degrees from the Faculty of Environmental Studies at the University of Waterloo following earlier studies in the School of Forestry at Lakehead University in Thunder Bay, Ontario. Dr. Franklin taught remote sensing and geographical information system courses at Memorial University of Newfoundland (1985–1988) and the University of Calgary (1988–2003) before joining the University of Saskatchewan in 2003 as Vice-President Research and Professor of Geography. He has served as associate editor of both the *Canadian Journal of Forest Research* (2000–2003) and the *Canadian Journal of Remote Sensing* (1991–1996) and has served on the editorial board of the journal *Computers and Geosciences* (1993–1996). Dr. Franklin is a Fellow of the Canadian Aeronautics and Space Institute and has served on the executive of the Canadian Remote Sensing Society, including two years as chair. His major publications include more than 100 journal articles in a wide range of remote sensing and environmental journals and the books *Remote Sensing for Sustainable Forest Management* (2001, Lewis/CRC) and *Remote Sensing of Forest Environments: Concepts and Case Studies* (2003, KAP). Dr. Franklin has supervised more than 30 graduate students and postdoctoral fellows in an environmental remote sensing research program, with applications in forest ecology and wildlife management, largely funded by the Natural Sciences and Engineering Research Council of Canada.

Contributors

Eric Arsenault
Canadian Forest Service (Northern
  Forestry Centre)
Natural Resources Canada
Edmonton, Alberta, Canada

Matthew Betts
Greater Fundy Ecosystem Research
  Group
University of New Brunswick
Fredericton, New Brunswick, Canada

Tom Bobbe
USDA Forest Service
Remote Sensing Applications Center
Salt Lake City, Utah

Mike Bobbitt
Department of Forest Resources
University of Idaho
Moscow, Idaho

Jess Clark
USDA Forest Service
Remote Sensing Applications Center
Salt Lake City, Utah

Warren B. Cohen
USDA Forest Service
Pacific Northwest Research Station
Corvallis, Oregon

Nicholas C. Coops
Department of Forest Resource
  Management
University of British Columbia
Vancouver, British Columbia, Canada

Steven E. Franklin
Department of Geography
University of Saskatchewan
Saskatoon, Saskatchewan, Canada

Sarah E. Gergel
Department of Forest Sciences and
  Centre for Applied Conservation
  Research
University of British Columbia
Vancouver, British Columbia, Canada

Ronald J. Hall
Canadian Forest Service (Northern
  Forestry Centre)
Natural Resources Canada
Edmonton, Alberta, Canada

Sean Healey
USDA Forest Service
Rocky Mountain Research Station
Ogden, Utah

Andrew Hudak
USDA Forest Service
Rocky Mountain Research Station
Moscow, Idaho

Robert Kennedy
USDA Forest Service
Pacific Northwest Research Station
Corvallis, Oregon

Michael B. Lavigne
Canadian Forest Service (Atlantic
  Forestry Centre)
Natural Resources Canada
Fredericton, New Brunswick, Canada

Leigh Lentile
Department of Forest Resources
University of Idaho
Moscow, Idaho

Julia Linke
Department of Geography
University of Calgary
Calgary, Alberta, Canada

Jennifer Miller
Department of Geology and Geography
West Virginia University
Morgantown, West Virginia

Penelope Morgan
Department of Forest Resources
University of Idaho
Moscow, Idaho

John Rogan
Graduate School of Geography
Clark University
Worcester, Massachusetts

Rob Skakun
Canadian Forest Service (Northern
 Forestry Centre)
Natural Resources Canada
Edmonton, Alberta, Canada

Joanne C. White
Canadian Forest Service (Pacific
 Forestry Centre)
Natural Resources Canada
Victoria, British Columbia, Canada

Michael A. Wulder
Canadian Forest Service (Pacific
 Forestry Centre)
Natural Resources Canada
Victoria, British Columbia, Canada

Yang Zhiqiang
Department of Forest Science
Oregon State University
Corvallis, Oregon

Contents

Michael A. Wulder and Steven E. Franklin

1 Introduction: Structure, Function, and Change of Forest Landscapes

Julia Linke, Matthew G. Betts, Michael B. Lavigne, and Steven E. Franklin

CONTENTS

INTRODUCTION

Forests are inherently dynamic in space and time. Their composition and distribution can change not only through continuous, subtle, and slow forest development and succession, but also through discontinuous, occasional, and sudden natural disturbances (Botkin, 1990; Oliver and Larson, 1996; Spies, 1997). In addition to natural processes, human activities and disturbances are the source of much contemporary forest change (Houghton, 1994; Meyer and Turner, 1994; Riitters et al., 2002). Such land cover change is widely considered the primary cause of biodiversity decline and species endangerment (Hansen et al., 2001). Monitoring natural and human-caused land cover and forest changes, disturbance processes, and spatial pattern is relevant for the conservation of forest landscapes and their inhabitants (Balmford et al., 2003). In recent years, international political momentum dedicated to conservation of biodiversity and sustainable development has increased (Table 1.1).

Biodiversity conservation and sustainable forest management require the collection of new kinds of forest and land cover information to complement traditional forest databases, model outputs, and field observations. Remote sensing and geographical information systems (GISs) have emerged as key geospatial tools — together with models of all kinds and descriptions — to satisfy increasing information needs of resource managers (Franklin, 2001). But, these are more than tools — they represent essentially new *approaches* to forest disturbance and spatial pattern mapping and analysis because they enable new ways of viewing disturbances and landscapes, which in turn influence our understanding and management practices. Critical developments in the use of remote sensing and GIS approaches include the ability to map biophysical (e.g., Iverson et al., 1989), biochemical (e.g., Roberts et al., 2003), and disturbance (e.g., Gong and Xu, 2003) characteristics of forest landscapes over a wide range of spatial scales and time intervals (Quattrochi and Pellier, 1991; Turner et al., 2003).

This introductory chapter provides a brief landscape ecological foundation for the importance of detecting and monitoring forest disturbances and changes in forest landscape patterns. We discuss monitoring and scale considerations and then describe basic stand and landscape dynamics of interest to resource managers. We introduce landscape metrics, which are then more completely reviewed by Gergel (Chapter 7, this volume). We emphasize a developing understanding of pattern/process reciprocity in forested landscapes, which is then highlighted by several case studies of different disturbance patterns in widely differing forest environments. Immediately following this introduction is background material on pertinent remote sensing and GIS data selection, methods, and applications issues in support of forest pattern analysis and change detection (Chapter 2). This material leads naturally to the suite of illustrative examples of remote sensing and GIS approaches in forest harvest pattern detection (Chapter 3), forest insect defoliation mapping (Chapter 4), monitoring fire disturbance (Chapter 5), and the role of GIS in forest disturbance and change mapping (Chapter 6). Subsequent chapters in this book present specific aspects of spatial pattern analysis, including remote sensing considerations (Chapter 7) and a detailed remote sensing/GIS/pattern analysis case study (Chapter 8) designed to aid in understanding critical resource management issues. Each of these chapters has been selected as a representative perspective on developing remote sensing and GIS approaches, which are increasingly recognized, in combination with field data and modeling methods, as the only feasible way to monitor landscape change over large areas with sufficient spatial detail to allow comparison of resultant patterns of different management or natural disturbance regimes.

TABLE 1.1
Selected National Programs on the Conservation of Biological Diversity and Sustainable Management of Earth Resources Developed Since the Rio Earth Summit, the United Nations Conference on Environment and Development (UNCED), in 1992

Program	Initiation Year and Organization	Vision	Web Site Address
Convention on Biological Diversity (CBD)	1992, United Nations Environment Programme	International treaty to pursue the conservation of biological diversity, the sustainable use of its components, and the fair and equitable sharing of the benefits arising out of the utilization of genetic resources, including by appropriate access to genetic resources and by appropriate transfer of relevant technologies, taking into account all rights over those resources and to technologies, and by appropriate funding	http://www.biodiv.org/default.shtml
The Montreal Process	1994, Inter-Governmental Organization of Forestry Agencies	A working group for the development of criteria and indicators that provide member countries with a common definition of what characterizes sustainable management of temperate and boreal forests	http://www.mpci.org/home_e.html
Pan-European Biological and Landscape Diversity Strategy	1994, Council of Europe	The principal aim of the strategy is to find a consistent response to the decline of biological and landscape diversity in Europe and to ensure the sustainability of the natural environment	http://www.coe.int/t/e/Cultural_Co-operation/Environment/Nature_and_biological_diversity/Biodiversity/default.asp#TopOfPage
Kyoto Protocol	1997, United Nations Framework Convention on Climate Change	Convention on Climate Change sets an overall framework for intergovernmental efforts to tackle the challenge posed by climate change with the Kyoto Protocol committing parties to individual, legally binding targets to limit or reduce their greenhouse emissions	http://unfccc.int/2860.php

Continued.

TABLE 1.1 (*Continued*)
Selected National Programs on the Conservation of Biological Diversity and Sustainable Management of Earth Resources Developed Since the Rio Earth Summit, the United Nations Conference on Environment and Development (UNCED), in 1992

Program	Initiation Year and Organization	Vision	Web Site Address
Earth Observation for Sustainable Development of Forests (EOSD)	1999, Canadian Forest Service and Canadian Space Agency	Research and applications program to develop a forest measuring and monitoring system that responds to key policy drivers related to climate change and to report on sustainable forest development of Canada's forest both nationally and internationally; space-based earth observation technologies are used to create products for forest inventory, forest carbon accounting, monitoring sustainable development, and landscape management	http://eosd.cfs.nrcan.gc.ca/
Earth System Science Partnership (ESSP)	2001, DIVERSITAS, International Geosphere-Biosphere Programme (IGBP), International Human Dimensions Programme on Global Environmental Change (IHDP), World Climate Research Programme (WCRP)	The ESSP brings together researchers from diverse fields and from across the globe to undertake an integrated study of the Earth system: its structure and functioning; the changes occurring to the system; and the implications of those changes for global sustainability	http://www.essp.org/
Millennium Ecosystem Assessment (MA)	2001, United Nations	International work program designed to meet the needs of decision makers and the public for scientific information concerning the consequences of ecosystem change for human well-being and options for responding to those changes	http://www.millenniumassessment.org//en/index.aspx

LANDSCAPE ECOLOGY

The traditional focus of forest ecology, management, and planning has been primarily on separate landscape elements such as homogeneous forest stands or habitat patches. The importance of interactions among different elements in a landscape was noted in the early 1980s (Forman, 1981), coincident with the need for forest management strategies to consider landscape structure as a requirement for long-term conservation of biodiversity (Noss, 1983; Risser et al., 1984). It has since become generally accepted that the structure of the landscape influences the ecological processes and functions that are operating within it (Haines-Young and Chopping, 1996). The discipline of *landscape ecology* is now widely recognized as a distinct perspective in resource management and ecological science.

The central goal of landscape ecology is the investigation of the reciprocal effects and interactions of landscape patterns and ecological processes (Turner, 1989). Fundamental to such investigation is the awareness that landscape observation is scale dependent, spatially and temporally, with different landscape patterns and processes discernible from different points of view and time that are specific to the organism (e.g., trees vs. earthworms) or the abiotic process (e.g., carbon gas fluxes) under study (Perera and Euler, 2000). A brief overview of general scale considerations is included in this introductory section; Coops et al. (Chapter 2, this volume) present concrete spatial data selection issues related to scale.

LANDSCAPE STRUCTURE, FUNCTION, AND CHANGE

When studying the ecology of landscapes, at least three basic elements must be considered and understood: structure, function, and change (Forman, 1995; M. Turner, 1989). Landscape *structure* generally refers to the distribution of energy, material, and species. The spatial relationships of landscape elements are characterized as landscape pattern in two ways (McGarigal and Marks, 1995; Remmel and Csillag, 2003). First, the simple number and amount of different spatial elements within a landscape is generally defined as landscape composition, and this measure is generally considered to be spatially implicit. Second, the arrangement, position, shape, and orientation of spatial elements within a landscape are generally defined as landscape configuration, which is a spatially explicit measure. Within the framework of this book, this meaning of landscape pattern is used to ensure that both the amount and arrangement of spatial elements of interest are included. In contrast, some studies equate landscape pattern strictly with configuration and treat composition as a second landscape characteristic unrelated to pattern (e.g., Martin and McComb, 2002; Miller et al., 2004).

A landscape can be defined as a spatially complex, heterogeneous mosaic in which homogeneous spatial elements or patches are repeated in similar form over an area bounded by the spatial scale at which ecological processes occur (Urban et al., 1987). For example, juvenile dispersal distance has been used to estimate the spatial extent of landscapes in forest birds (Villard et al., 1995); in another example, a third-order watershed could be the appropriate landscape for consideration of water flow and quality (Betts et al., 2002). Mosaic patterns exist at all spatial scales from submicroscopic to the planet and universe and the type, size, shape, boundary, and arrangement of landscape elements across this mosaic influence a variety of ecological functions.

Landscape *function* generally refers to the flow of energy, materials, and species and the interactions between the mosaic elements (Forman, 1995). Examples range from fundamental abiotic processes, such as cycling of water, carbon, and minerals (Waring and Running, 1998), to biotic processes, including forest succession (Oliver and Larson, 1996), and the dispersal and gene flow of wildlife (e.g., Hansson, 1991). Such biotic and abiotic flows are determined by the landscape structures present, and in turn, landscape structure is created and changed by these flows. The main processes or flows generating landscape structure formation and landscape *change* over time can be considered as natural and anthropogenic disturbances (e.g., wildfire, insect infestation, harvesting); biotic processes (e.g., succession, birth, death, and dispersal); and environmental conditions (e.g., soil quality, terrain, climate) (Levin, 1978). An overview of some of these processes in the forest environment is presented in a subsequent section of this chapter and in later chapters discussing specific disturbance processes.

FOREST MANAGEMENT

The goals of forest management have expanded in recent decades to include values leading to the implementation of different strategies based on concepts of sustained yield, multiple use, and more recently, ecosystem management. Ecosystem management includes the balancing ecological and social (economic and noneconomic) forest values in the context of increasing population growth, resource use, pollution, and the rate and extent of ecosystem alteration (Kimmins, 2004). Concepts of natural disturbance emulation encompass the idea of trying to arrange changes in forests due to human disturbance to more closely approximate those induced by natural processes (Attiwill, 1994; Hunter, 1990). This is an acknowledgment of disturbances as one of the fundamental processes and drivers of landscape structure and functioning at all spatial and temporal scales in the field of landscape ecology (Turner, 1987). Principles of landscape ecology help to make this forest management approach a viable management option by providing a higher-level context for forest management practices (Crow and Perera, 2004).

Emulating natural disturbance aims to guide local forest management by mimicking the natural range of spatial and temporal variation in landscape- and stand-level forest landscape structures created by past natural disturbances in the given location (Bergeron et al., 1999; Hunter, 1999; Kimmins, 2004). The presettlement landscape allowing for natural dynamism is thought to be the ideal condition against which contemporary landscape diversity and composition ought to be evaluated (Noss, 1983; Seymour and Hunter, 1999). The natural disturbance approach builds on the underlying assumption that forest ecosystems, long-term forest stability, and biodiversity will be sustained if the forest structures created by natural disturbances are maintained since they reflect the same conditions under which these ecosystems have evolved (Bunnell, 1995; Engelmark et al., 1993; Hunter, 1990). For example, Hudak et al. (Chapter 8, this volume) provide a case study perspective of forest harvest and fire disturbance patterns in an area where both disturbances are known to have occurred.

Consideration of the ecological effects of spatial patterns created by forest harvesting is important for the management regime (Franklin and Forman, 1987), and the patterns and processes in landscapes created by natural disturbances generally display greater variation in time and space than traditional silviculture and forest management (Seymour et al., 2002). Disturbance regimes can be described by a variety of characteristics; however, the main components include magnitude, timing, and spatial distribution (Seymour and Hunter, 1999), and each of these will have an impact on the stand- (or patch-) and landscape-level of the forest ecosystem. Magnitude generally describes the intensity or the physical force of the disturbance or the severity of the effect of the disturbance on the landscape element or organism (Seymour and Hunter, 1999; Turner et al., 2001). Timing of a disturbance mainly specifies the frequency, which is often expressed not only as the return interval between disturbances, but also as the duration and seasonality of a disturbance type (Seymour and Hunter, 1999). The spatial distribution of a disturbance refers to the extent, shape, and arrangement of disturbance patches (Seymour and Hunter, 1999).

A review by Seymour et al. (2002) of disturbance regimes in northeastern North America contrasts the differences in aspects of these three main characteristics (magnitude, timing, and spatial distribution) by comparing wildfire with pathogens and insect herbivory. In the investigated cases, wildfires were of stand-replacing magnitude, with a return interval of 806 to 9000 years and a disturbance patch size distribution ranging between 2 and more than 80,000 ha, while pathogens and insect herbivory disturbance was of a magnitude to create smaller canopy gaps, with a return interval and patch size distribution ranging between 50 and 200 years and between 0.0004 and 0.1135 ha, respectively (Seymour et al., 2002; Figure 1.1).

While the natural disturbance approach may be an ecologically sound premise, its constraints and limitations also need to be considered. Some issues to address in the future include (a) society's reluctance to accept this paradigm in ecosystems that experience disturbances that are very large, severe, and frequent; (b) whether past disturbance regime effects will be rendered inapplicable in the future due to long-term climatic variation, invasion of nonnative species, air pollution, human-induced climate change (Kimmins, 2004); and (c) the difficulty in obtaining and interpreting historic disturbance data for adequate conclusions about the natural disturbance characteristics (Appleton and Keeton, 1999).

SCALE

Every organism is an "observer" of the environment, and every observer looks at the world through a filter, imposing a perceptual bias that influences the recognition of natural systems (Levin, 1992). Science, in general, can be seen as a product of the way the world is seen, constrained by the space and time within which humans inhabit the world (Church, 1996). There is little doubt that ecologists' perceptions have been revolutionized through availability of satellite imagery; for example:

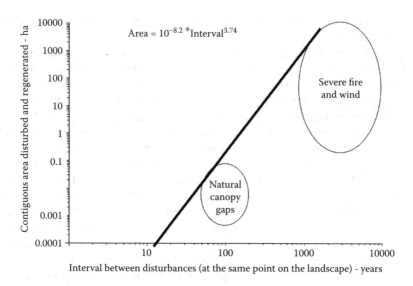

FIGURE 1.1 Boundaries of natural variation in studies of disturbance in northeastern North American forests. The hand-fitted diagonal boundary line defines the upper limits on these disturbance parameters in combination, all of which fall in the lower right of the diagram. Upper limits of the area and return interval of severe fires and windstorms were truncated at 10^4 Ha and 10^4 years, respectively. (Adapted from Seymour et al., 2002.)

- "Images from satellites have revolutionized our perception and approaches to understanding landscapes and regions" (Forman, 1995: p. 35)
- "More than any other factor, it was this perspective provided by satellite imagery that changed the ... manager's views about the main threats to the panda's survival" (Mackinnon and de Wulf, 1994, p. 130)

Scale is a strong determinant of viewing, and interpreting the environment and the interest in scale-related research is rapidly increasing (Schneider, 1994). Scale is often understood simply as dimensions of time and space, but has been defined in various more complex ways; for example, Church (1996) considered scale as a relative measure set by the resolution of measurements. Schneider (1994, p.3) defined scale as "the resolution within the range of a measured quantity." Common to all scientific definitions of scale, however, is a recognition of the temporal and spatial dimensions (Lillesand and Kiefer, 2000; Wiens, 1989).

SPATIAL SCALE

In ecology, spatial scale is usually considered as the product of *grain* and *extent* (Forman, 1995; Wiens, 1989), which, in remote sensing, relate to *resolution* (pixel size) and *area of coverage,* respectively (Lillesand and Kiefer, 2000). A remote sensing scientist will typically define spatial scale as a proportion, a ratio of length on a map to actual length. Small scale, therefore, suggests that a large area is covered; in other words, the difference between actual and mapping size is great (coarse spatial detail). An ecologist's typical definition of spatial scale is the level or degree of spatial resolution and spatial extent perceived or considered. Ecologists understand a small-scale study to encompass a small area with fine spatial detail. Overall extent and grain define the upper and lower limits of resolution of a study; they are analogous to the overall size of a sieve and its mesh size (Wiens, 1989). The spatial scale at which measurements or observations are taken influences the recognition of spatial patterns and underlying processes of the environment and of the organisms under study (Wiens, 1989); this has been called *intrinsic* scale, which may determine the type of spatial patterns observed. "The intrinsic scale is a property of the ecological process of interest, for example, tree fall, competition, stomatal control, or microclimate feedbacks, and it is governed in part by the size of the individual organisms (or events) and in part by the range of their interactions with their environment" (Malingreau and Belward, 1992, p. 2291). Others (e.g., Hunsaker et al., 2001) have been keen to understand the *uncertainty* associated with spatial data at different scales.

Remotely sensed imagery is an optimal way to collect spatial data across multiple nested or hierarchical scales; imagery can provide synoptic coverage over large areas, enabling investigations at the landscape scale, or more detailed imagery can be collected representing smaller areas, most practically through some form of sampling framework. As always, limitations exist in the quantities of spatial resolution and area of coverage that can be obtained. Spatial resolution of imagery depends on the sensor spectral sensitivity, and the instantaneous field of view, while the area of coverage depends on the satellite or airborne altitude (swath width) and the instrument total field of view (Lillesand and Kiefer, 2000; Richards and Jia, 1999). Landsat satellites typically cover an area of 185 × 185 km with a sensor spatial resolution or pixel size of 30 × 30 m for most of the spectral bands; other satellites carrying Advanced Very High Resolution Radiometer (AVHHR) sensors cover an area of 2394 × 2394 km with a spatial resolution of approximately 1.1 km. More details on these fundamental concepts are presented in Chapter 2 of this volume.

TEMPORAL SCALE

Temporal scale refers to the frequency with which an observation is made (Lillesand and Kiefer, 2000), but similar to the spatial scale, it is made up of two components; the temporal resolution and the temporal extent. The key to temporal scale is change over time, and this pattern or trend may change with hours, days, months, years, or centuries. Depending on the research question and the object under study, the temporal scale of the investigation can be very different. For each source of imagery, the temporal resolution — a sensor-specific component of scale — must be quantified. Satellites passing frequently over the same area translates into a higher temporal resolution for a given sensor package; for example, the temporal resolution is 24 days for Indian Resource Satellite (IRS)–P2 satellites (Richards and Jia, 1999), but 1 day for satellites carrying the AVHRR (Malingreau and Belward, 1992). In addition, the original start of data collection for different sensor packages determines the maximum possible temporal extent of any earth observation study. Operable satellites launched many years ago translate into a higher temporal extent; for example, the

IRS-P2 satellite was launched in October 1994 (Richards and Jia, 1999), while AVHRR satellites were launched in several National Oceanic and Atmospheric Administration series between June 1979 and May 1991. Clearly, the ability to monitor frequent landscape changes at the temporal scale desired (e.g., daily) may be limited by the temporal resolution and extent of a given satellite platform.

RESEARCH DESIGN AND INTERPRETATION

Understanding the effect of scale on the detection and understanding of patterns and causal mechanisms is one step toward the development of common ecological theories within scales (Wiens, 1989). There is no single proper scale at which all sampling ought to be undertaken (Levin, 1992; Wiens, 1989), and there are no simple rules to select automatically the appropriate scales of attention (Meentemeyer, 1989). Ecological structure, function, and change are dependent on spatial and temporal scale (Turner, 1989). The identification of the appropriate scale to use will depend on the organism or phenomenon under investigation. A species- or phenomenon-centered approach, with recognition of its intrinsic scale to the identification of structure, is most relevant in the research design and analysis of forest landscapes.

Arbitrary scale choices can be avoided by analyzing the variance of measurements across many scales using techniques such as the nearest neighbor method (Davis et al., 2000), semivariance analysis (Meisel and Turner, 1998), and several other univariate (spatial correlograms and spectral analysis) and multivariate methods (Mantel test and Mantel correlogram; Legendre and Fortin, 1989). Statistical approaches are typically based on the observation that variance increases as transitions are approached in hierarchical systems (O'Neill et al., 1986). Peaks of unusually high variance indicate scales at which the between-group differences are especially large, which suggest the representation of the scale of natural aggregation or patchiness of vegetation (Greig-Smith, 1952) or organisms; this is sometimes referred to as the boundary of a scale domain (Wiens, 1989). A method of identifying the appropriate scale of remotely sensed imagery uses a high spatial resolution image characterized statistically and then subsequently collapsed to successively coarser spatial resolutions while calculating local variance (Woodcock and Strahler, 1987). The image resolution at which local variance is highest can be deemed the appropriate remote sensing scale in relation to the structural components of the ground.

PROCESSES GENERATE PATTERNS

Remote sensing of terrestrial ecosystems in support of resource management involves identifying ecosystems and their biological, ecological, and physical characteristics (Franklin, 2001). The definition of an ecosystem and the relevant characteristics vary with the resource managed and the issue under consideration. Therefore, the expectations that ecologists might have of remote sensing will vary; for example, species composition and the physical arrangement of the vegetation can be remotely sensed and used to describe or infer ecosystem attributes using straightforward methods and readily available data. Advances in remote sensing technology continue to expand the capacity to monitor changes of interest in ecosystems and resource management (Wulder et al., 2004). Forest ecosystems change over time because the trees must grow to survive, due to competition among trees, interactions among trophic levels, and large-scale disturbances. Certain aspects of the current state of ecosystem dynamism can be inferred from individual, remotely sensed images, and other aspects can only be assessed using a time series of images. In this section, we provide ecological background on the remote sensing of ecosystem attributes with special attention to the dynamic nature of these ecosystem attributes, the landscape structure, and composition.

FOREST STAND DYNAMICS

Current understanding of patterns and processes of stand development have been fully described by Oliver and Larson (1996). Their synthesis is useful as a basis for understanding the potential contributions of remote sensing. *Disturbance*, meaning the death of trees that frees growing space, is fundamentally important for stand development. Oliver and Larson (1996) distinguished between autogenic and allogenic forms of disturbance; *autogenic* processes cause death of individual trees for reasons that are particular to the tree and ecosystem, and *allogenic* forms of disturbance arise outside of the affected trees or ecosystem. For ease of explaining the processes involved in stand dynamics and the stand structures that result, Oliver and Larson first described long-term stand development following a major disturbance, including autogenic processes responsible for death of trees, and then incorporated the impacts allogenic forms of disturbance imposed on this underlying pattern of stand development. Oliver and Larson pointed out that stand development has been investigated from two perspectives, one based on describing stand structures and the other based on understanding stand developmental processes. The latter approach has great value to resource management because it leads to greater capacity for predicting changes to stands over time. Individual remotely sensed images may be well suited to the stand structural approach to understanding stand dynamics, while stand development typically requires multitemporal resolution imagery. Ecological knowledge must be used to interpret the remotely sensed images to ensure maximum information extraction occurs from available remotely sensed data (Graetz, 1990).

Forest ecosystems pass through four stages during the course of stand development (Figure 1.2). The period immediately following a major disturbance is the stand initiation stage. During this stage, the important process in stand dynamics is the establishment of a cohort of vegetation. New vegetation becomes established when the preexisting vegetation is killed; the number of species and the number of plants that establish themselves and grow to fill the unoccupied growing space depends on the ecoclimatic zone, site capacity to supply essential materials (nutrients and water), and the relative amount of growing space that is made available and the manner in which it is made available. The period of recruitment ends when the community of trees first comes to fully occupy the available growing space. At this time, the ecosystem enters the stem exclusion stage. Competition among established trees is the dominant process affecting ecosystem development and structure during the stem exclusion stage. Inherent differences among species affect the course of competition and consequently the stand structures that develop. Virtually no growing

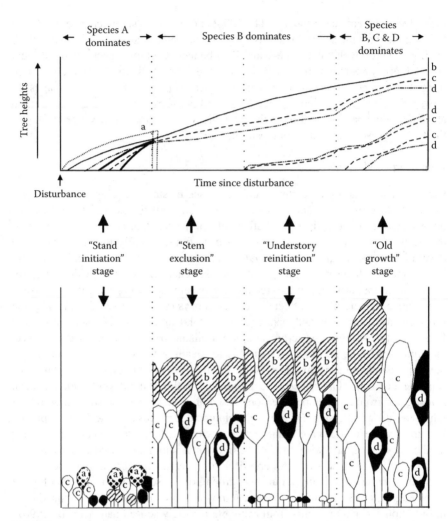

FIGURE 1.2 Schematic stages of stand development following major disturbances. All trees forming the forest start soon after the disturbance; however, the dominant tree type changes as stem number decreases and vertical stratification of species progresses. The height attained and the time lapsed during each stage vary with species, disturbance, and site. (Adapted from Oliver and Larson, 1996.)

space becomes available for the establishment of additional trees as the result of density-dependent mortality (competition). At about the time that the height growth of successful competitors becomes negligible, these trees begin losing their ability to maintain their "grip" on the growing space. This diminished capacity might be abetted by disease or the activities of insects commonly found in the ecosystem and eventually some trees die.

Species that have been less successful in competing in previous years may now expand to fill the vacated growing space and consequently come to dominate the overstory. However, if some of the growing space that comes available is captured by a ground story, particularly a ground story that includes advanced regeneration of some tree species, then the stand enters the understory reinitiation stage. Stand structure becomes increasingly complex with the onset of the understory reinitiation stage; the advanced regeneration does not have sufficient growing space to form a lower strata of the canopy. Consequently, the ecosystem remains dominated by the cohort of trees that were established after the initial disturbance. At a later time in stand development, the autogenic processes release growing space in sufficiently large areas to cause patches to return to the stand initiation stage, and as a result the ecosystem enters the old-growth stage. With the release of growing space in these patches, advanced growth is released, and other regeneration mechanisms operate to cause a new cohort of trees to become established. The establishment of patches of vegetation of new cohorts continues until all of the original cohort has been replaced, and at this time an old-growth stand exists. In nature, this stage of development is seldom achieved because in many parts of the globe large-scale disturbances return the entire ecosystem to the stand initiation stage. Other forests are influenced by gap-replacing disturbance, and there continues to be considerable debate about the historical frequency of gap versus stand-replacing disturbances. One possible valuable application of remote sensing would be to test some of the assumptions about the frequency and extent of gap versus stand-replacing forest disturbances (Wulder et al., 2004).

Oliver and Larson (1996) presented that it is more common for a variety of tree species ranging from pioneers to long-lived, shade-tolerant species to become established during the stand initiation stage (known as *initial floristics*) than for later seral stage species to become established after early seral stage species have occupied the site, modified the environment, and lived a substantial portion of their life cycle (*relay floristics*). This is in contrast to ideas of early ecologists, who imagined that stand development involved a succession of stand cover types. Moreover, Oliver and Larson (1996) show that forest ecosystems commonly develop a stratified mixed stand structure during the stand initiation and stem exclusion stages. In stratified mixed stands, the pioneer species grow most rapidly in the years immediately following a disturbance and dominate the overstory in the years immediately following the disturbance. Species with inherently slow initial height growth but capable of surviving in shade, albeit with even slower growth rates, sort themselves into lower strata during the early years of the stem exclusion stage. Species that initially dominate the upper stratum are usually shorter lived than the more tolerant species in lower strata, and hence eventually the lower strata are freed from suppression and dominate the overstory. The difference between the initial floristic pattern of stand establishment and the relay floristic pattern has practical significance when interpreting the pattern of stand development of stratified mixed stands. In the past, stratified mixed stands have been sometimes misinterpreted to be uneven-aged stands, whereas in reality members of all strata became established in response to the same disturbance. This distinction is particularly important when devising silvicultural interventions to maintain or promote particular stand structures.

Site characteristics such as microclimate and soil conditions vary spatially, affecting the mix of species that becomes established in the various ecosystems that make up a landscape. During the stand initiation stage, site characteristics can be viewed as environmental "sieves" through which species must pass to become established. For example, species vary in their capacity to tolerate drought, grow on nutrient-poor soils, become established on cold sites, withstand exposure, and survive in shade. Many remotely sensed images only contain information about the uppermost canopy layer and not about lower strata and the ground story, but knowledge of ecological habits of the tree species and of the stand development patterns operating in the region can be used to better interpret current stages in stand development of the observed ecosystems and their future stand structures. Some promising new image data types with three-dimensional capabilities (e.g., light detection and ranging, LIDAR) are described by Coops et al. (Chapter 2, this volume).

EFFECTS OF DISTURBANCE ON STAND DYNAMICS AND LANDSCAPES

Fire, windthrow, insect attack, and harvesting are examples of allogenic disturbances. Each type of disturbance has a different impact on ecosystems and landscapes, thereby having diverse effects on the stand structure created, the species that can become established in the growing space made available by the disturbance, and changes to the soil and site conditions necessary for tree growth. The frequency and spatial extent of major disturbances affect the proportion of a landscape in each stage of stand development at any point in time. Remote sensing provides data for monitoring disturbances and documenting their effects on each ecosystem in a landscape. These data can provide a means to monitor the subsequent stand development for much larger numbers of ecosystems than could be measured by field surveys.

The type and severity of disturbance affects the success that can be achieved by each regeneration strategy. For example, forest fires commonly consume the forest floor, thereby eliminating advanced regeneration and therefore some species such as balsam fir, which rely on advanced growth to become established after disturbance and are prevented from being a future part of the ecosystem after fire. Clark and Bobbe (Chapter 5, this volume) provide background and an example of using remote sensing for portraying fire impacts, and Hudak et al. (Chapter 8, this volume) include fire disturbance in the presented case study. A contrasting example on the role of disturbance in favoring particular regeneration strategies is what happens when the disturbance removes selected species from the overstory but does not eliminate the ground story, such as occurs with outbreaks of defoliating insects (Seymour, 1992; Figure 1.3). In these instances, species regenerating from advanced growth might have an advantage in acquiring the growing space made available, and species that are intolerant of shade might be limited in their capacity to regenerate. Hall et al. (Chapter 4, this volume) develop an approach useful in mapping insect disturbance using remote sensing.

Disturbances can create atypical stand structures in ecosystems that had been in the stem exclusion or understory reinitiation stages at the time of disturbance. These relatively young stands would typically have complete overstories of trees belonging to one cohort if only allogenic processes were in play, but after a major disturbance that does not completely eliminate the original cohort, these stands will have more than one cohort visible from above. Such disturbed stands will exhibit greater spatial variability in stand structure than undisturbed stands. It is possible that these disturbed stands will have structure commonly associated with old-growth stands, and these structures might mislead some into believing that they are stable, old-growth

Virgin stand ca. 1860

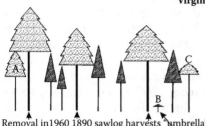

△ Red spruce
🌲 Balsam fir

Large, old red spruce dominate the overstory of the multi-aged stand after enduring long periods of suppression on lower strata. Earlier developmental stages are illustrated by the smaller (and probably younger) spruces A, B, and C in the present stand. Firs generally occupy sub-canopy strata.

Removal in 1960   1890 sawlog harvests   "umbrella" spruce

Typical post-harvest structure ca. 1890

Repeated sawlog harvests remove old spruce. Residual overstory now consists of previously suppressed, unmerchantable spruces and firs. Advance seedlings and saplings begin to devolop in gaps created by harvesting.

Before spruce budworm outbreak ca. 1910

Many residual firs respond vigorously, and now dominate the overstory. Residual spruces also respond; some trees over 15 cm dbh are removed for pulpwood during early 1900s. Advance regeneration forms dense, virtually continuous lower stratum with an irregular structure corresponding to the tree heights when released.

After spruce budworm outbreak ca. 1925

All mature firs (and some spruces) killed by budworm outbreak ca. 1913–1919. Many advance fir saplings also succumb; some survive but suffer severe dieback of their terminal shoots.

Second-growth stand ca. 1970

After recovery, stand quickly returns to stem exclusion stage, as the sapling advance regeneration develops vigorously. Where remnant spruces B and C (from the original old-growth stand) do not blow down or succumb to bark beetles, they now dominate the overstory. Where budworm-caused mortality and logging removed the overstory completely, stand has very even-aged structure, with fir generally dominant over spruce.

FIGURE 1.3 Development of typical spruce-fir stand after logging and budworm attack circa 1860–1970. (Adapted from Seymour 1992.)

ecosystems. Disturbance can also change the capacity of a site to supply the nutrients and water required for establishment and growth. These effects can increase or decrease growth, and the impact on growth might vary over time. For example, fire can release nutrients held in recalcitrant organic matter, thereby increasing plant growth immediately after the fire, but fire also decreases the total stock of nitrogen existing on the site, which may decrease long-term potential productivity.

A disturbance affecting a large area can reduce the number of tree species that can disperse seeds onto the disturbed areas. Dispersal distance varies among species depending on mode of dispersal, seed size, and special appendages on seeds that facilitate dispersal. Therefore, species such as trembling aspen with very light seeds that can disperse in wind for great distances can disperse seeds onto large disturbed areas, whereas species with heavy seeds and no wings or other appendages to facilitate dispersal can only disperse a short distance from their site of origin (Burns and Honkala, 1990). In some circumstances, species with restricted dispersal distances maintain a presence in stands that become established because variations in the severity of the disturbance leave islands of living trees throughout the disturbed area to serve as seed sources. For example, small patches of burned or lightly burned forest are sometimes found scattered across a burned landscape, and seeds from species such as white spruce can disperse from these refugia into the surrounding disturbed area.

Large-scale disturbance is an integral part of numerous landscapes affecting the species found and the distribution of land area among stages of development. As a consequence, large-area forest fires affect virtually all of the boreal forest of central Canada on a regular basis (Stocks et al., 2003). Only tree species that are adapted to regenerate after fire are found in this region. Moreover, most stands are in the stem exclusion stage or early in the understory reinitiation stages because the fire frequency precludes many stands reaching the old-growth stage. Similarly, widespread outbreaks of spruce budworm in the spruce-fir forests of eastern North America profoundly affect stand development and the landscape (Baskerville, 1975). Spruce budworm outbreaks occur at 30-year intervals, and in northern New Brunswick there were two age classes of ecosystems existing at the beginning of the outbreak: 30-year-old forest stands originating from the immediately previous outbreak and 60-year-old stands originating from the outbreak before the last one (Figure 1.4). There is widespread mortality in the 60-year-old stands, causing those ecosystems to return to the stand initiation stage. In the 30-year-old stands, the effect of spruce budworm mortality is to thin stands and shift species composition to a higher percentage of spruce and birch and a lower percentage of balsam fir, although balsam fir usually continues to be the largest proportion of these stands. In this manner, disturbance has largely determined the landscape characteristics and significantly affected stand development of most stands in the region.

PATTERNS GENERATE PROCESSES

IMPACTS OF PATTERNS ON ECOLOGICAL PROCESSES

Just as physical and ecological processes generate landscape structure, landscape structure influences physical and ecological processes. Specifically, landscape pattern

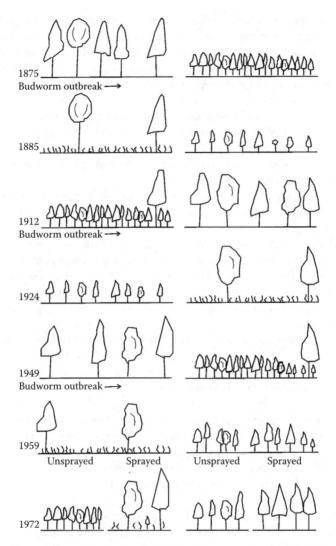

FIGURE 1.4 Natural succession in the Green River fir-spruce-birch forest. The two columns represent two sequences both beginning in 1875 but with differing initial conditions. The blackened tops represent disfiguration by intensive budworm feeding. The hatched cones represent white spruce trees, the unshaded cones are firs, and the rounded trees are hardwoods. (Adapted from Baskerville, 1975.)

has been found to affect rates of wind and water erosion, intensity of natural disturbances (Foster, 1988), plant and animal movement (Beier and Noss, 1998), survival (Doherty and Grubb, 2002), and reproduction (Robinson et al., 1995). Here, a brief review is provided of the components of landscape pattern that have been shown to exert a strong influence on ecological processes. Such patterns are considered priorities for measurement in remote sensing (Gergel, Chapter 7, this volume).

All landscapes are characterized by degrees of heterogeneity (patchiness) at different scales; differing substrates (soils, bedrock), natural disturbances (fire, insect outbreaks), and human activity (forestry, road building) all create patchiness across a landscape. The "patch-corridor-matrix" model (Forman, 1995) has become a central component of landscape ecology in theory and in practice:

1. A *patch* is a homogenous area that differs from its surroundings (Forman, 1995). A woodlot surrounded by farmland and a wetland immersed in upland habitat are examples of patches. Patch shape often correlates with the intensity of human activity. Intense human activity often results in simpler, less-convoluted patch shape

2. *Corridors* are a form of patch in that they differ from the surrounding areas. However, they are usually identified as strips that aid in flows between patches (Lindenmayer, 1994). Corridors fulfill a number of roles, including facilitating animal dispersal, wildlife habitat, preventing soil and wind erosion, and aiding in integrated pest management (Barrett and Bohlen, 1991). A riparian buffer strip might serve as a corridor for forest songbirds (Machtans et al., 1996) or a kilometers-wide forested strip could serve as a corridor for cougar (Beier, 1995). The life history traits of each species determine the characteristics of corridor habitat

3. The *matrix* is the most extensive component of the landscape, is highly connected, and controls regional dynamics (Forman, 1995). For instance, in the Canadian prairies, small woodlot patches occur in a matrix of natural grassland or agricultural development

The landscape structures briefly described above (patches, corridors, and matrix) influence, and are influenced by, landscape flows. These flows include diverse elements such as wildlife (Lindenmayer and Nix, 1993), soil and nutrients (Stanley and Arp, 1998), and water (Campbell, 1970). For example, Haddad (1999) demonstrated that pine plantations impede the movement of butterflies between patches of early successional forest.

One of the central principles of landscape ecology is that all ecosystems are interrelated, with movement or flow rate dropping sharply with distance but more gradually between ecosystems of the same type (Forman, 1995). Thus, a very heterogeneous landscape (with many patch types) is marked by a relatively low degree of movement (flow) and a large amount of resistance.

FRAGMENTATION, CONNECTIVITY, AND ISOLATION

Fragmentation is the "breaking apart" of habitat. This can occur as a result of natural processes such as forest fires or anthropogenic disturbances such as road building or timber harvesting. Different views exist about definitions regarding "habitat loss" and "fragmentation." Wilcove et al. (1986) suggested that fragmentation is a combination of habitat loss and isolation; however, recently the emerging consensus is that habitat loss and fragmentation should be described separately (Andrén, 1994; Fahrig, 1998, 2002; Mazerolle and Villard, 1999). Fragmentation is often defined

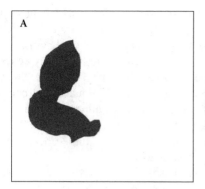
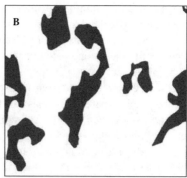

FIGURE 1.5 Unfragmented (A) versus fragmented (B) landscapes with the same amount of habitat present in each landscape.

purely as the breaking apart of habitat and does not always imply habitat loss. For instance, holding habitat area constant, a landscape can either be fragmented (i.e., many patches) or unfragmented (i.e., one patch) (Figure 1.5). While habitat loss and fragmentation are often confounded in real landscapes (i.e., they occur together), we emphasize that it is important to determine which of these is ecologically important; if populations respond to habitat fragmentation, land managers may be able to design landscapes that mitigate risks.

Landscape fragmentation effects may be grouped into a few major categories, including edge, patch size, and distance between patches (connectivity) (Schmiegelow and Mönkkönen, 2002). For example, edges are the result of vegetation boundaries in the landscape and may result from (a) enduring features (soils, drainage, slope); (b) natural disturbances; or (c) human activities such as forest harvesting or farm development. An edge effect may be caused by differences in moisture, temperature, and light that occur along the boundary between different adjacent patch types (Saunders et al., 1991). A number of studies have reported increased rates of predation at forest edges (Paten, 1994); however, this appears to be context dependent (Batáry and Báldi, 2004). Many organisms are affected by the size of favorable habitat patches. Such species are termed *area sensitive* (Freemark and Collins, 1992). Robbins et al. (1989) found that "area" was one of the most significant habitat features for many neotropical migrant bird species. Area sensitivity has also been observed for amphibians (Hager, 1998). Although some debate exists about the area sensitivity of plants, a number of published studies reported lower genetic diversity and higher rates of extinction in smaller populations (Bell et al., 1991; Damman and Cain, 1998).

In some cases, fragmented landscapes have been shown to exhibit the same characteristics as those observed in island archipelagos by MacArthur and Wilson (1967). Isolation of habitat seems to compound the effect of small patch size on the ability of some species to persist and recolonize. These findings can be understood better if placed in the context of the concept of *metapopulations*. The metapopulation concept requires that population dynamics be studied beyond the scale of local populations. "Equilibrium," rather than occurring in a single, local population, might occur as a result of a number of interconnected subpopulations that are distributed across a region (Husband and Barrett, 1996). Population dynamics are the result of a series of local extinctions and recolonizations in habitat patches (Levins, 1970). If the subpopulation of one patch becomes extinct, then it may eventually be recolonized by dispersers from a subpopulation that exists in a neighboring patch. This is the "rescue effect"; for a species to spread or persist, individuals must colonize unoccupied habitat patches at least as frequently as populations become extinct (Hanski and Ovaskainen, 2000). As fragmentation progresses, the distance between patches (isolation) of mature forest increases. This distance limits the ability of organisms to disperse and colonize new habitat patches.

It is important to note that, in addition to the studies briefly described above indicating a significant influence of landscape pattern on species distributions, there are many studies that reveal only weak or nonexistent landscape pattern effects (Delin and Andrén, 1999; Game and Peterken, 1984; McGarigal and McComb, 1995; Schmiegelow et al., 1997; Simberloff and Gotelli, 1984). Indeed, the majority of evidence indicates that it is habitat loss rather than fragmentation per se that is the most important influence on species occurrence, reproduction, and survival (Fahrig, 2003). This appears particularly to be the case in forest mosaics (for reviews, see Bender et al., 1998; Mönkkönen and Reunanen, 1999). This idea reinforces the notion that it is important for remote sensing to provide accurate classifications of landscape composition in addition to input data to analyses of landscape configuration.

Andrén (1994) proposed that landscape configuration is only important below a threshold in the proportion of suitable habitat at the landscape scale. Only at low levels of habitat are patches small and isolated enough to result in patch size effects or restrictions in movement (Gardner et al., 1991). This results in multiplicative impacts of fragmentation on habitat loss. A number of theoretical studies supported this "fragmentation threshold" hypothesis (Fahrig, 1998; Hill and Caswell, 1999; Wiegand et al., 2005; With and King, 1999), but it has rarely been demonstrated in nature (Trzcinski et al., 1999). However, this may be because "suitable habitat" has rarely been defined according to the requirements of individual species.

PREDICTIVE MODELING OF SPECIES OCCURRENCE USING GEOSPATIAL DATA

To determine rates of change in the amount and pattern of habitat at any spatial scale, it is clearly necessary to have accurate definitions of habitat for different species. Remotely sensed data have been used extensively to develop habitat models; these are inexpensive to develop in comparison to models based on detailed vegetation data collected in the field (Osborne et al., 2001; Vernier et al., 2002) and provide an opportunity to generate predictions about species distributions over large spatial extents at relatively fine resolutions (Betts et al., 2006; Gibson et al., 2004; Linke et al., 2005). Such models are usually probabilistic in nature, but a wide range of modeling techniques are available, including classification trees, neural networks, generalized linear models, generalized additive models, and spatial interpolators (Segurado and Araujo, 2004). Models have been developed to cover aspects as diverse as biogeography, conservation biology, climate change research, habitat or species management (Guisan and Zimmermann, 2000), and vegetation mapping (J.

F. Franklin, 1995). As the resolution of remotely sensed data improves, the range of potential applications is likely to increase (Coops et al., Chapter 2, this volume; Wulder et al., 2004).

LANDSCAPE METRICS

To study the effects of landscape structure on ecological processes, it is necessary to develop methods to quantify spatial patterns into measurable variables before links to ecological processes can be determined (Frohn, 1998). *Landscape metrics*, or indices, have been developed to meet this need (Diaz, 1996). Early landscape metric studies presented only a few metrics, typically *dominance* (the degree to which certain kinds of landscape patches or classes predominate in the landscape), *contagion* (the extent to which similar patches are clumped together), and *shape* (the form of an area or a patch as determined by the variation of its border) (Forman, 1995; O'Neill et al., 1988). Today, an extensive array or suite of landscape metrics and indices exists (Elkie et al., 1999; McGarigal and Marks, 1995). The suite of available landscape metrics can be considered to include specific measures of area, edge, shape, core (or interior) area, nearest neighbor/diversity/richness/evenness, interspersion/juxtaposition, contagion/configuration, and connectivity/circuitry (Gergel, Chapter 7, this volume; McGarigal et al., 2000).

The large number of metrics that have been developed to describe and quantify spatial structure often appears to be overwhelming, and the question of metric redundancy has frequently arisen. Initially, use of metrics that have known ecological relevance and application should be considered. However, some standard approaches have been employed to deal with the issues of redundancy and number of metrics for a given application. For example, Riitters et al. (1995) used a factor analysis to reduce to a few components more than 50 specific landscape metrics applied to 85 maps of land use and cover in the United States. Recent studies have concluded that it is possible to identify a parsimonious suite of metrics using principal components analysis to characterize much of the spatial patterns existing in a boreal forest landscape subject to many common disturbance processes (S. Cushman, personal communication, April and October, 2002; Linke and Franklin, in press).

In addition to the issue of appropriate metric selection, there are several other factors known potentially to influence the interpretation and use of landscape metrics (Haines-Young and Chopping, 1996), including, for example, metric uniqueness, sensitivity, abrupt versus continuous edges, statistical quantification, study area extent, and scale or resolution. Another important characteristic of landscape metrics to consider is their actual behavior over a wide range of landscape structures; the instance of nonlinear landscape metric behavior over scale is briefly mentioned here:

1. Hargis et al. (1998) investigated the relationships between six landscape metrics and the proportion of two landcover types across simulated land-scapes, also controlling for the size and shape of patches. Most metrics were linearly associated at the lower landcover proportion range but had nonlinear associations at higher proportions, which limits their direct comparability across different regions

2. Such nonlinear metric behavior was also found in simulated landscapes in a study of dispersal success on fractal landscapes (With and King, 1999) and in a study specifically designed to detect metric behavior under controlled conditions (Neel et al., 2004)

An awareness of all of these interpretational effects and metric behavioral limitations must be embedded in any landscape quantification attempt. A detailed discussion of spatial pattern analysis using landscape metrics is presented by Gergel in Chapter 7.

CONCLUSION

Understanding forest disturbance and spatial pattern is increasingly recognized as essential to effective and sustainable forest management in many forest environments around the world. We hope that this introduction has provided some insight into the challenges that are further elaborated in later chapters; the developing appetite in landscape ecology and conservation biology for spatial data and models that work with complex phenomena; the relationships between pattern and process, process and pattern; the specific details of remotely sensed and GIS data selection; the importance of scale; the myriad issues in fire and insect, forest harvesting, and other disturbance monitoring; and the emerging role of landscape metrics and modeling landscapes. The literature and practice of forest disturbance and spatial pattern using remote sensing and GIS approaches are diverse and increasing at an astonishing rate as new perspectives and insights take hold. We expect this presentation will be useful to those involved in this interesting and exciting endeavor, in the implementation and continued development of remote sensing and GIS approaches, and in their application to forest ecosystems and processes. We anticipate progress in these areas will help shape future directions in the important work of forest resource mapping, monitoring, and management.

ACKNOWLEDGMENTS

The support of the Canadian Forest Service and the Natural Sciences and Engineering Research Council of Canada is gratefully acknowledged. We acknowledge the careful review and many helpful suggestions to improve this chapter provided during the review process.

REFERENCES

Andrén, H. (1994). Effects of habitat fragmentation on birds and mammals in landscapes with different proportions of suitable habitat: a review. *Oikos, 71,* 355–366.

Appleton, G.H. and Keeton, W.S. (1999). Applications of historic range of variability concepts to biodiversity conservation. In R.K. Baydack, H. Champa III, and J.B. Haufler (Eds.), *Practical Approaches to the Conservation of Biological Diversity* (pp. 71–86). Island Press, Washington, DC 313p.

Attiwill, P.M. (1994). The disturbance of forest ecosystems: the ecological basis for conservative management. *Forest Ecology and Management, 63*, 247–300.

Balmford, A., Green, R.E., and Jenkins, M. (2003). Measuring the changing state of nature. *Trends in Ecology and Evolution, 18(7)*, 326–330.

Barrett, G. and Bohlen, P. (1991). Landscape ecology. In W. Hudson (Ed.), *Landscape Linkages and Biodiversity* (pp. 149–161). Island Press, Washington, DC. 194p.

Baskerville, G.L. (1975). Spruce budworm: super silviculturist. *The Forestry Chronicle, August,* 138–140.

Batáry, P. and Báldi, A. (2004). Evidence of an edge effect on avian nest success. *Conservation Biology, 18,* 389–400.

Beier, P. (1995). Dispersal of juvenile cougars in fragmented habitat. *Journal of Wildlife Management, 59,* 41–67.

Beier, P. and Noss, R.F. (1998). Do habitat corridors provide connectivity? *Conservation Biology, 12,* 1241–1252.

Bell, G., Martin, J., and Schoen, D. (1991). The ecology and genetics of fitness in forest plants III. Environmental variance in natural populations of *Impatiens pallida. Journal of Ecology, 79,* 697–713.

Bender, D.J., Contreras, T.A., and Fahrig, L. (1998). Habitat loss and population decline: a meta-analysis of the patch size effect. *Ecology, 79,* 517–533.

Bergeron, Y., Harvey, B., Leduc, A., and Gauthier, S. (1999). Forest management guidelines based on natural disturbance dynamics: stand and forest-level considerations. *Forestry Chronicle, 75,* 49–54.

Betts, M., Forbes, G., and Knox, J. (2002). A landscape ecological approach to private woodlot planning in New Brunswick. *Natural Areas Journal, 22,* 311–317.

Betts, M.G., Diamond, A.W., Forbes, G.J., Villard, M.-A., and Gunn, J. (2006). The importance of spatial autocorrelation, extent and resolution in predicting forest bird occurrence. *Ecological Modelling, 191,* 197–224.

Botkin, D.B. (1990). *Discordant Harmonies: A New Ecology for the Twenty-First Century.* Oxford University Press, New York. 244p.

Bunnell, F.L. (1995). Forest-dwelling vertebrate faunas and natural fire regimes in British Columbia: patterns and implications for conservation. *Conservation Biology, 9,* 636–644.

Burns, R.M. and Honkala, B.H. (1990). *Silvics of North America. Volume 2, Hardwoods.* U.S. Department of Agriculture, Forest Service, Washington, DC. Agriculture Handbook 654. 877p.

Campbell, C. (1970). Ecological implications of riparian vegetation management. *Journal of Soil and Water Conservation, 2,* 49–52.

Church, M. (1996). Space, time and the mountain — how do we order what we see? The scientific nature of geomorphology. *Proceedings of the 27th Binghamton Symposium in geomorphology* (pp. 147–170). Edited by Rhoads, B.L. and C.E. Thorn. John Wiley & Sons, New York.

Crow, T.R. and Perera, A.H. (2004). Emulating natural landscape disturbance in forest management — an introduction. *Landscape Ecology, 19,* 231–233.

Damman, H. and Cain, M. (1998). Population growth and viability analyses of the clonal woodland herb, *Asarum canadense. Ecology, 86,* 13–26.

Davis, J.H., Howe, R.W., and Davis, G.J. (2000). A multi-scale spatial analysis method for point data. *Landscape Ecology, 15,* 99–114.

Diaz, N.M (1996). Landscape metrics. A new tool for forest ecologists. *Journal of Forestry, 94(12),* 12–16.

Delin, A.E. and Andrén, H. (1999). Effects of habitat fragmentation on Eurasian red squirrel (*Sciurus vulgaris*) in a forest landscape. *Landscape Ecology, 14,* 67–72.

Doherty, P.F. and Grubb, T.C. (2002). Survivorship of permanent-resident birds in a fragmented forested landscape. *Ecology, 83*, 844–857.

Elkie, P.C., Rempel, R.S., and Carr, A.P. (1999). *Patch Analyst User's Manual: A Tool for Quantifying Landscape Structure.* Ministry of Natural Resources, Northwest Science and technology, Thunder Bay, ON, Canada. NWST Technical Manual TM-002. 22p.

Engelmark, O., Bradshaw, R., and Bergeron, Y. (1993). Disturbance dynamics in boreal forests: introduction. *Journal of Vegetation Science, 4*, 730–732.

Fahrig, L. (1998). When does fragmentation of breeding habitat affect population survival? *Ecological Modelling, 105*, 273–292.

Fahrig, L. (2002). Effect of habitat fragmentation on the extinction threshold: a synthesis. *Ecological Applications, 12*, 346–353.

Fahrig, L. (2003). Effects of habitat fragmentation on biodiversity. *Annual Reviews of Ecology and Systematics, 34*, 487–515.

Forman, R.T.T. (1981). Interaction among landscape elements: a core of landscape ecology. In S.P. Tjallingii and A.A. de Veer (Eds.), *Perspectives in Landscape Ecology* (pp. 35–48). PUDOC, Wageningen, Netherlands. 344p.

Forman, R.T.T. (1995). *Land Mosaics, the Ecology of Landscapes and Regions.* Cambridge University Press, Cambridge, U.K. 632p.

Foster, D.R. (1988). Species and stand response to catastrophic wind in central New England. *U.S.A. Journal of Ecology, 80*, 753–772.

Franklin, J.F. (1995). Predictive vegetation mapping: geographic modelling of biospatial patterns in relation to environmental gradients. *Progress in Physical Geography, 19*, 474–499.

Franklin, J.F. and Forman, R.T.T. (1987). Creating landscape patterns by forest cutting: ecological consequences and principles. *Landscape Ecology, 1*, 5–18.

Franklin, S. E. (2001). *Remote Sensing for Sustainable Forest Management,* Lewis Publishers/CRC Press, Boca Raton, FL. 407p.

Freemark, K. and Collins, B. (1992). Landscape ecology of birds breeding in temperate forest fragments. In J. Hagan and D. Johnson (Eds.), *Ecology and Conservation of Neotropical Migrant Landbirds* (pp. 443–454). Smithsonian Institution Press, Washington, DC. 609p.

Frohn, R.C. (1998). *Remote Sensing for Landscape Ecology: New Metric Indicators for Monitoring, Modeling, and Assessment of Ecosystems.* Lewis Publishers, New York. 99p.

Game, M. and Peterken, G. (1984). Nature reserve selection strategies in the woodlands of central Lincolnshire, England. *Biological Conservation, 29*, 157–181.

Gardner, R.H., Turner, M.G., O'Neill, R.V., and Lavorel, S. (1991). Simulation of scale-dependent effects of landscape boundaries on species persistence and dispersal. In M.M. Holland, P.G. Risser, and R.J. Naiman (Eds.), *Ecotones: The Role of Landscape Boundaries in the Management and Restoration of Changing Environments* (pp. 76–89). Chapman and Hall, New York. 142p.

Gibson, L.A., Wilson, B.A., Cahill, D.M., and Hill, J. (2004). Spatial prediction of rufous bristlebird habitat in a coastal heathland: a GIS-based approach. *Journal of Applied Ecology, 41*, 213–223.

Gong, P. and B. Xu, B.(2003). Remote sensing of forests over time: change types, methods, and opportunities. In M.A. Wulder and S. E. Franklin (Eds.), *Remote Sensing of Forest Environments: Concepts and Case Studies* (pp. 301–333). Kluwer Academic Publishers, Norwell, MA. 519p.

Graetz, R.D. (1990). Remote sensing of terrestrial ecosystem structure: An ecologist's pragmatic view. In R.J. Hobbs and H.A. Mooney (Eds.), *Remote Sensing of Biosphere Functioning* (pp. 5–30). Springer-Verlag, New York. 312p.

Greig-Smith, P. (1952). The use of random and contiguous quadrats in the study of the structure of plant communities. *Annals of Botany, 16*, 293–316.

Guisan, A. and Zimmermann, N.E. (2000). Predictive habitat distribution models in ecology. *Ecological Modelling, 135*, 147–186.

Haddad, N.M. (1999). Corridor and distance effects on interpatch movements: a landscape experiment with butterflies. *Ecological Applications, 9*, 612–622.

Hager, H.A. (1998). Area-sensitivity of reptiles and amphibians: are there indicator species for habitat fragmentation? *Ecoscience, 5*, 139–147.

Hansen, A.J., Neilson, R.P., Dale, V.H., Flather, C.H., Iverson, L.R., Currie, D.J., Shafer, S., Cook, R., and Bartlein, P.J. (2001). Global change in forests: responses of species, communities, and biomes. *Bioscience, 51*, 765–779.

Hanski, I. and Ovaskainen, O. (2000). The metapopulation capacity of a fragmented landscape. *Nature, 404*, 755–758.

Hansson, L. (1991). Dispersal and connectivity in metapopulations. *Biological Journal of the Linnean Society, 42*, 89–103.

Haines-Young, R. and Chopping, M. (1996). Quantifying landscape structure: a review of landscape indices and their application to forested landscapes. *Progress in Physical Geography, 20*, 418–445.

Hargis, C.D., Bissonette, J.A., and David, J.L. (1998). The behaviour of landscape metrics commonly used in the study of habitat fragmentation. *Landscape Ecology, 13*, 167–186.

Hill, M.F. and Caswell, H. (1999). Habitat fragmentation and extinction thresholds on fractal landscapes. *Ecology Letters, 2*, 121–127.

Houghton, R.A. (1994). The worldwide extent of land-use change. *Bioscience, 44*, 305–313.

Hunsaker, C., Goodchild, M., Friedl, M., and Case, E., (Eds.), (2001). *Spatial Uncertainty in Ecology: Implications for Remote Sensing and GIS Applications.* Springer-Verlag, New York. 402p.

Hunter, M.L., Jr. (1990). *Wildlife, Forests, and Forestry: Principles of Managing Forests for Biological Diversity.* Prentice-Hall, Englewood Cliffs, NJ. 370p.

Hunter, M.L., Jr. (1999). *Maintaining Biodiversity in Forest Ecosystems.* Cambridge University Press, Cambridge, U.K. 698p.

Husband, B.C. and Barrett, S.C.H. (1996). A metapopulation perspective in plant biology. *Journal of Ecology, 84*, 461–469.

Iverson, L.R., Graham, R.L., and Cook, E. A. (1989). Applications of satellite remote sensing to forested ecosystems. *Landscape Ecology, 3*, 131–143.

Kimmins, J.P.H. (2004). Emulating natural forest disturbance: what does this mean? In A.H. Perera, L.J. Buse, and M.G. Weber (Eds.), *Emulating Natural Forest Landscape Disturbances: Concepts and Applications* (pp. 8–28). Columbia University Press, New York. 315p.

Legendre, P. and Fortin, M.J. (1989). Spatial pattern and ecological analysis. *Vegetatio, 80*, 107–138.

Levin, S.A. (1978). Pattern formation in ecological communities. In J.S. Steele (Ed.), *Spatial Pattern in Plankton Communities* (pp. 433–465). Plenum Press, New York. 482p.

Levin, S.A. (1992). The problem of patterns and scale in ecology. *Ecology, 73*, 1943–1967.

Levins, R. (1970). Extinction. In M. Gerstenhaber (Ed.), *Some Mathematical Questions in Biology* (pp. 77–107). American Mathematical Society, Providence, RI.

Lillesand, T.M. and Kiefer, R.W. (2000). *Remote Sensing and Image Interpretation.* 4th ed. John Wiley & Sons, New York. 724p.

Lindenmayer, D. (1994). Wildlife corridors and the mitigation of logging impacts on fauna in wood-production forests in south-eastern Australia: a review. *Wildlife Research, 21,* 323–340.

Lindenmayer, D. and Nix, H. (1993). Ecological principles for the design of wildlife corridors. *Conservation Biology, 7,* 627–630.

Linke, J. and Franklin, S.E. (In press). Interpretation of landscape metric gradients based on satellite image classification of land cover. *Canadian Journal of Remote Sensing.*

Linke, J., Franklin, S.E., Huettmann, F., and Stenhouse, G. (2005). Seismic cutlines, changing landscape metrics and grizzly bear landscape use in Alberta. *Landscape Ecology, 20,* 811–826.

Machtans, C., Villard, M.-A., and Hannon, S. (1996). Use of riparian buffer strips as movement corridors by forest birds. *Conservation Biology, 10,* 1366–1379.

Mackinnon, J. and de Wulf, R. (1994). Designing protected areas for giant pandas in China. In R.I. Miller (Ed.). *Mapping the Diversity of Nature* (pp.127–142). Chapman and Hall, London.

Malingreau, J.P. and Belward, A.S. (1992). Scale considerations in vegetation using AVHRR data. *International Journal of Remote Sensing, 13,* 2289–2307.

Martin, K.J. and McComb, W.C. (2002). Small mammal habitat associations at patch and landscape scales in Oregon. *Forest Science, 48,* 255–264.

Mazerolle, M. and Villard, M-A. (1999). Patch characteristics and landscape context as predators of species presence and abundance: a review. *Ecoscience, 6,* 117–124.

MacArthur, R. and Wilson, E.O. (1967). *The Theory of Island Biogeography.* Princeton University, Princeton, NJ. 203p.

McGarigal, K. and Marks, B.J. (1995). *FRAGSTATS. Spatial Pattern Analysis Program for Quantifying Landscape Structure.* Reference Manual. Forest Science Department Oregon State University, Corvallis, OR. 122p.

McGarigal, K. and McComb, W. (1995). Relationships between landscape structure and breeding birds in the Oregon coast range. *Ecological Monographs, 65,* 235–260.

McGarigal, K., Cushman, S., and Stafford, S. (2000). *Multivariate Statistics for Wildlife and Ecology Research.* Springer-Verlag, New York. 283p.

Meentemeyer, V. (1989). Geographical perspectives of space, time, and scale. *Landscape Ecology, 3,* 163–173.

Meisel, J.E. and Turner, M.G. (1998). Scale detection in real and artificial landscapes using semivariance analysis. *Landscape Ecology, 13,* 347–362.

Meyer, W.B. and Turner, B.L. II. (1994). *Changes in Land Use and Land Cover: A Global Perspective.* Cambridge University Press, Cambridge, U.K. 537p.

Miller, J.R., Dixon, M.D., and Turner, M.G. (2004). Response of avian communities in large-river floodplains to environmental variation at multiple scales. *Ecological Applications, 14,* 1394–1410.

Mönkkönen, M. and Reunanen, P. (1999). On critical thresholds in landscape connectivity: a management perspective. *Oikos, 84,* 302–305.

Neel, M., McGarigal, K., and Cushmann, S.A. (2004). Behaviour of class-level landscape metrics across gradients of class aggregation and area. *Landscape Ecology, 19,* 435–455.

Noss, R.F. (1983). A regional landscape approach to maintain diversity. *BioScience, 33,* 700–706.

Oliver, C.D. and Larson, B.C. (1996). *Forest Stand Dynamics.* John Wiley & Sons, New York. 520p.

O'Neill, R.V., DeAngelis, D.L., Waide, J.B., and Allen, T.F.H. (1986). *A Hierarchical Concept of Ecosystems*. Princeton University Press, Princeton, NJ. 253p.

O'Neill, R.V., Krummel, J.R., Gardner, R.H., Sugihara, G., Jackson, B., DeAngelis, D.L., Milne, B.T., Turner, M.G., Zygmunt, B., Christensen, S.W., Dale, V.H., and Graham, R.L. (1988). Indices of landscape pattern. *Landscape Ecology, 1*, 153–162.

Osborne, P.E., Alonso, J.C., and Bryant, R.G. (2001). Modelling landscape-scale habitat use using GIS and remote sensing: a case study with Great Bustards. *Journal of Applied Ecology, 38*, 458–471.

Paten, P.W.C. (1994). The effect of edge on avian nest success: how strong is the evidence? *Conservation Biology, 8*, 17–26.

Perera, A.H. and Euler, D.L. (2000). Landscape ecology in forest management: An introduction. In A.H. Perera, D.L. Euler, and I.D. Thompson (Eds.), *Ecology of a Managed Terrestrial Landscape: Patterns and Processes of Forest Landscapes in Ontario* (pp. 3–11). UBC Press, Vancouver, BC, Canada. 336p.

Quattrochi, D.A. and Pellier, R.E. (1991). Remote sensing for analysis of landscapes: an introduction. In M.G. Turner and R.H. Gardner (Eds.), *Quantitative Methods in Landscape Ecology: The Analysis and Interpretation of Landscape Heterogeneity* (pp. 51–76). Springer-Verlag, New York. 536p.

Remmel, T.K. and Csillag, F. (2003). When are two landscape pattern indices significantly different? *Journal of Geographical Systems, 5*, 331–351.

Richards, J.A. and Jia, X. (1999*). Remote Sensing Digital Image Analysis*. Springer, New York. 363p.

Riitters, K.H., O'Neill, R.V., Hunsaker, C.T., Wickham, J.D., Yankee, D.H., Timmins, S.P.,Jones, K.B., and Jackson, B.L. (1995). A factor analysis of landscape pattern and structure metrics. *Landscape Ecology, 10*, 23–39.

Riitters, K.H., Wickham, J.D., O'Neill, R.V., Jones, K.B., Smith, E.R., Coulston, J.W., Wade, T.G., and Smith, J.H. (2002). Fragmentation of continental United States forests. *Ecosystems, 5*, 815–822.

Risser, P.G., Karr, J.R., and Forman, R.T.T. (1984). Landscape ecology: directions and approaches. *Illinois Natural History Survey Special Publication No. 2*. Illinois Natural History Survey, Champaign, IL. 18p.

Robbins, C.S., Dawson, D., and Dowell, B. (1989). Habitat area requirements of breeding forest birds of the Middle Atlantic States. *Wildlife Monographs, 103*, 1–34.

Roberts, D.A., Keller M., and Soares, J.V. (2003). Studies of land-cover, land-use, and biophysical properties of vegetation in the Large Scale Biosphere Atmosphere experiment in Amazonia. *Remote Sensing of Environment, 87*, 377–388.

Robinson, S.K., Thompson, F.R., Donovan, T.M., Whitehead, D.R., and Faaborg, J. (1995). Regional forest fragmentation and the nesting success of migratory birds. *Science, 267*, 1987–1989.

Saunders, D., Hobbs, R., and Margules, C. (1991). Biological consequences of ecosystem fragmentation: a review. *Conservation Biology, 5*, 18–32.

Schmiegelow, F.K.A., Machtans, C., and Hannon, S. (1997). Are boreal birds resilient to forest fragmentation? An experimental study of short-term community responses. *Ecology, 78*, 1914–1932.

Schmiegelow, F.K.A. and Mönkkönen, M. (2002). Habitat loss and fragmentation in dynamic landscapes: avian perspectives from the boreal forest. *Ecological Applications, 12*, 375–389.

Schneider, D.C. (1994). *Quantitative Ecology, Spatial and Temporal Scaling*. Academic Press, New York. 395p.

Segurado, P. and Araujo, M.G. (2004). An evaluation of methods for modelling species distributions. *Journal of Biogeography, 31*, 1555–1568.

Seymour, R.S. (1992). The red spruce — balsam fir forest of Maine: evolution of silvicultural practice in response to stand development patterns and disturbances. In M.J. Kelty, B.C. Larson, and O.D. Oliver (Eds.), *The Ecology and Silviculture of Mixed-Species Forests* (pp. 217–244). Kluwer Academic Publishers, Dordrecht, Netherlands. 287p.

Seymour, R.S. and Hunter, M.L., Jr. (1999). Principles of ecological forestry. In M.L. Hunter Jr. (Ed.), *Maintaining Biodiversity in Forest Ecosystems* (pp. 22–64). Cambridge University Press, Cambridge, U.K. 698p.

Seymour, R.S., White, A.S., and deMaynadier, P.G. (2002). Natural disturbance regimes in northeastern North America — evaluating silvicultural systems using natural scales and frequencies. *Forest Ecology and Management, 155*, 357–367.

Simberloff, D. and Gotelli, N. (1984). Effects of insularisation on plant species richness in the prairie-forest ecotone. *Biological Conservation, 29*, 27–46.

Spies, T.A. (1997). Forest stand structure, composition and function. In K.A. Kohm and J.F. Franklin (Eds.), *Creating a Forestry for the 21st Century, the Science of Ecosystem Management* (pp. 11–30). Island Press, London.

Stanley, B. and Arp, P. (1998). *Timber Harvesting of Forested Watershed: Impacts on Stream Discharge*. Faculty of Forestry and Environmental Management, Fredericton, NB, Canada.

Stocks, B.J., Mason, J.A., Todd, J.B., Bosch, E.M., Wotton, B.M., Amiro, B.D., Flannigan, M.D., Hirsch, K.G., Logan, K.A., Martell, D.L., and Skinner, W.R. (2003). Large forest fires in Canada, 1959–1997. *Journal of Geophysical Research. Atmospheres. 108 (D1) 8149*, doi:10.1029/2001JD000484.

Trzcinski, M.K., Fahrig, L., and Merriam, G. (1999). Independent effects of forest cover and fragmentation on the distribution of forest breeding birds. *Ecological Applications, 9*, 586–593.

Turner, M.G. (1987). *Landscape Heterogeneity and Disturbance*. Springer Verlag, New York. 241 p.

Turner, M.G. (1989). Landscape ecology: the effect of pattern on process. *Annual Review of Ecology and Systematics, 20*, 171–197.

Turner, M.G., Gardner, R.H., and O'Neill, R.V. (2001). *Landscape Ecology in Theory and Practice — Pattern and Process*. Springer-Verlag, New York. 352p.

Turner, W., Spector, S., Gardiner, N., Fladeland, M., Sterling, E., and Steininger, M. (2003). Remote sensing for biodiversity science and conservation. *Trends in Ecology and Evolution, 18*, 306–314.

Urban, D.L., O'Neill, R.V., and Shugart, H.H., Jr. (1987). Landscape ecology. A hierarchical perspective can help scientists understand spatial patterns. *BioScience, 3*, 119–127.

Vernier, P., Schmiegelow, F.K.A., and Cumming, S. (2002). Modeling bird abundance from forest inventory data in the boreal mixed-wood forests of Canada. In J.M. Scott, P.J. Heglund, M.L. Morrison, J.B. Haufler, M.G. Raphael, W.A. Wall, and F.B. Sampson (Eds.), *Predicting Species Occurrences, Issues of Accuracy and Scale* (pp. 271–280). Island Press, Washington, DC. 868p.

Villard, M.-A., Merriam, G., and Maurer, B.A. (1995). Dynamics in subdividing populations of neotropical migratory birds in a fragmented temperate forest. *Ecology, 76*, 27–40.

Waring, R.H. and Running, S.W. (1998). *Forest Ecosystems. Analysis at Multiple Scales*. 2nd ed. Academic Press, San Diego, CA. 370p.

Wiegand, T., Revilla, E., and Moloney, K.A. (2005). Effects of habitat loss and fragmentation on population dynamics. *Conservation Biology, 19*, 108–121.

Wiens, J.A. (1989). Spatial scaling in ecology. *Functional Ecology, 3*, 385–397.

Wilcove, D.S., McLellan, C.H., and Dobson, A.P. (1986). Habitat fragmentation in temperate zones. In M.E. Soule (Ed.), *Conservation Biology: The Science of Scarcity and Diversity* (pp. 237–256). Sinauer Assoc., Sunderland, MA.

With, K.A. and King, A.W. (1999). Extinction thresholds for species in fractal landscapes. *Conservation Biology, 13*, 314–326.

Woodcock, C.E. and Strahler, A.H. (1987). The factor of scale in remote sensing. *Remote Sensing of Environment, 21*, 311–332.

Wulder, M.A., Hall, R.J., Coops, N., and Franklin, S.E. (2004). High spatial detail remotely-sensed data for ecosystem characterization. *Bioscience, 54(6),* 511–521.

2 Identifying and Describing Forest Disturbance and Spatial Pattern: Data Selection Issues and Methodological Implications

Nicholas C. Coops, Michael A. Wulder, and Joanne C. White

CONTENTS

INTRODUCTION

An increasing number of remotely sensed data sources are available for detecting and characterizing forest disturbance and spatial pattern. As the information that is extracted from remotely sensed data is often a function of image characteristics, matching the appropriate data source to the disturbance target of interest requires knowledge of these image characteristics. Furthermore, an understanding of the implications of the dependencies between imagery selected, disturbance of interest, and change detection approach followed is required to facilitate the selection of an appropriate data source. The method used to capture the disturbance information must also be considered within the context that not all methods inherently support all data sources and vice versa. The goals of this chapter are to identify the key issues for consideration during the data selection process; highlight how these issues have an impact on the successful detection and characterization of forest disturbance and spatial pattern; and finally review the range of methods available for detecting forest disturbances and emphasize the link between these methods and the selection of an appropriate data source.

BACKGROUND

Observations of ecological disturbances have been acquired since remote sensing technologies first became available (Cohen and Goward, 2004). Since the invention of photography, it was apparent that images captured from the air provided important information on the spatial patterns on the Earth's surface (Colwell, 1960) and quickly became critical for resource managers. As early as the 1910s, for example, barely a decade after the first aerial remote sensing platforms were developed, the synoptic view afforded by aerial sensors benefited a number of disciplines, including forestry and ecology (Spurr, 1948). During the 1920s, improved camera systems for producing vertical aerial photographs with minimal distortion were developed (Thompson and Gruner, 1980). As a result, the U.S. Department of Agriculture systematically began to photograph agricultural lands throughout the United States in the 1930s (Rango et al., 2002). By 1950, aerial photography was a standard tool for resource managers concerned with mapping land cover and land use change (Goward and Williams, 1997). Aerial coverage has continued to the present day to provide an invaluable resource to examine the dynamics of spatial pattern (Goslee et al., 2003; Rango et al., 2002).

Space-based remote sensing of the Earth's surface began from *Explorer 6* in 1959 and the Television Infra Red Observation Satellite (TIROS) National Oceanic and Atmospheric Administration (NOAA) series of satellites began in 1960 (Goward and Williams, 1997). Since then, imagery from the Advanced Very High Resolution Radiometer (AVHRR) and more recently from the Moderate Resolution Imaging

Spectroradiometer (MODIS) sensors on TERRA and AQUA has made routine mapping of global vegetation possible (Cohen et al., 2002; Running et al., 1999). However, these systems are principally designed for global coverage with low spatial resolution (approximately 1 to 5 km), which is generally too coarse for monitoring localized or regional disturbance events (Cohen et al., 2002). Imagery at much finer spatial scales, at around 80 m, has been available since the launch of Landsat-1 in 1972 (Cohen and Goward, 2004). Since then, a family of Landsat satellites have orbited the Earth, with many other similar successful satellite programs initiated by other countries, including France, India, Japan, and Russia (Stoney, 2004). Successful launches of both commercial and government satellite programs over the past five years (and those planned for the next five years) have resulted in a large increase in the number of available satellite-based imaging sensors. In 2005, there were expected to be up to 30 satellites with spatial resolutions ranging from 0.3 m to 2.5 km (Stoney, 2004).

As of the writing of this chapter, Landsat-5 is experiencing some technical difficulties, and Landsat-7 is not operating as envisioned, with a scan line corrector problem requiring the production of mosaicked image products. The status of the Landsat sensors, both operationally and politically, is changing rapidly. While continuity of the collection of Landsat-like data is enshrined as public policy, the continuity of the actual Landsat sensor program is currently not clear. Efforts are under way to ensure some form of data collection of Landsat-like data. The actual sensor, timing, and mechanisms for this to happen are currently not known. Current information can be found at http://landsat.usgs.gov/.

SELECTION OF REMOTELY SENSED IMAGERY

As discussed by Linke et al. (Chapter 1, this volume), mapping and monitoring landscape disturbance is highly scale dependent — both spatially and temporally. As a result, the landscape patterns and processes that are discernible from any particular remotely sensed image source are dependent on the target of interest (e.g., single tree versus stand-replacing disturbance) and the spatial, spectral, radiometric, and temporal characteristics of the image source (Perera and Euler, 2000; Turner, 1989). These image characteristics must be considered during the data selection process along with the methods and techniques that may be used to detect the change. In addition, the information requirements of the end user must be considered. For example, an end user interested in the total area disturbed by fire in a single year may be satisfied with a simple binary classification indicating areas of fire and no fire. Conversely, an end user interested in forest succession following a fire event may require more detailed information on fire extent, as well as species composition and abundance, to monitor the pattern of forest succession over time. Image characteristics will dictate which image source is most appropriate for the given information need.

The characteristics of a remotely sensed image are often collectively referred to as the image resolution and relate to the size of individual pixels or picture elements, the overall spatial extent of the image, the time interval of acquisition, the level of detail or discrimination the sensor is capable of providing, the region(s) of the electromagnetic spectrum in which the sensor collects data, and the bit depth of the sensor. Each of these image characteristics and the interactions between these image characteristics, are addressed in the following sections. In addition, the implications of these characteristics for data selection, in the context of forest disturbance, are discussed.

SPATIAL RESOLUTION

The spatial resolution of a remotely sensed scene provides an indication of the size of the minimum area that can be resolved by a detector at an instant in time (Strahler et al., 1986; Woodcock and Strahler, 1987). In the case of aerial photography, the spatial resolution is based on the film speed or size of the silver halide crystal (Nelson et al., 2001). In the case of digital sensors, an instrument that has a spatial resolution of 30 m is technically able to resolve any 30 m by 30 m area on the landscape as one single reflectance response. The information content of a pixel is tied to the relationship between the spatial resolution and the size of the objects of interest on the Earth's surface. If trees are the objects of interest and a sensor with a 30-m spatial resolution is used, then many objects (trees) per pixel can be expected, which limits the utility of the data for characterizing the individual trees. However, if forest stands are the objects of interest and an image source with a 30-m pixel is used, a number of pixels will represent each forest stand, resulting in an improved potential for characterization of stand-level attributes. This relationship between pixels and objects was fully characterized by Woodcock and Strahler (1987). The area that is covered by a single remotely sensed image (spatial extent or image footprint) is principally a function of the sensor swath width or field of view (Lillesand and Kiefer, 2000; Richards and Jia, 1999). Instruments with a low spatial resolution typically have the capacity to capture larger areas. For example, Landsat Thematic Mapper (TM; considered a medium spatial resolution sensor) images have a spatial extent of 185 by 185 km with a spatial resolution of 30 m (for most of its spectral bands). Conversely, the NOAA AVHHR sensor has a much larger swath width and subsequently covers a greater area (2394 by 2394 km) with a spatial resolution of between 1.1 and 6 km (Richards and Jia, 1999), with the range in spatial resolution due to off-nadir (i.e., not directly beneath the sensor, but at an angle) scanning during data capture.

The generalized description of spatial resolution indicates an expectation of the nature of the information that is captured (Woodcock and Strahler, 1987). High spatial resolution data may provide detailed information on objects as finite as individual trees, streams, or buildings; however, the image footprint or spatial extent is also typically limited (e.g., 10 by 10 km), often precluding use of this data for large area studies for both feasibility and cost reasons (Wulder, Hall, et al., 2004). Historically, medium spatial resolution sensors (such as Landsat TM and SPOT multispectral imagery) have provided the optimal resolution for characterizing large areas with comprehensive coverage while maintaining an ability to describe land-scape-level phenomena, such as land cover change and regional disturbance (Franklin and Wulder, 2002; Woodcock et al., 2001). Further, the nature of the patterns identified from high spatial resolution data differs from those captured from lower spatial resolution data (e.g., trees vs. stands). Gergel (Chapter 7, this volume) addresses issues related to the investigation of high spatial resolution data with landscape pattern metrics. Traditional trends in landscape pattern metrics found when analyzing Landsat or lower spatial resolution data will not necessarily be found when analyzing higher spatial resolution data as the patterns represent different surface or vegetation characteristics.

Figure 2.1 provides an example of how the information content of a remotely sensed image can vary with spatial resolution. Figure 2.1A is a Landsat-7 Enhanced Thematic Mapper Plus (ETM +) multispectral image representing an area of approximately 8 km^2. With the 30-m spatial resolution of the ETM+ data, broad-scale features such as forest stands, harvest blocks, and roads are discernible. A subarea representing approximately 0.5 km^2 is shown in Figure 2.1B and Figure 2.1C. Panel B is a portion of a multispectral QuickBird image with a spatial resolution of 2.44 m. At this spatial resolution, individual trees can be identified. In the Figure 2.1C, a portion of a digital aerial photograph with a 0.30-m pixel is shown; individual trees can be resolved with greater detail than in the QuickBird image, and furthermore, the attributes associated with these individual trees can be characterized. For example, trees damaged by mountain pine beetle appear red in the digital photo (note that the same area of red-attack damage is also present in the QuickBird image).

LIDAR (light detection and ranging) data represent the three-dimensional structure of the surface or vegetation canopy. LIDAR systems emit a pulse of laser infrared radiation and measure the time (and therefore distance) it takes for the pulse to reach and then be reflected by the surface (Lefsky and Cohen, 2003). LIDAR data are typically collected as single points or profiles; therefore, the land surface is sampled rather than fully imaged, resulting in noncontiguous data. Most airborne systems have a point spacing between 1 to 5 m depending on the system configuration and flying altitude and speed, which may be customized to meet user needs (Lim et al., 2003). These points are in turn processed to represent ground and canopy elevation surfaces.

When selecting a data source for forest disturbance mapping, spatial resolution will be a key decision. Table 2.1 outlines the optimal applications associated with different spatial resolutions. Generally, broad-scale phenomena (covering large areas, for which general trends are of interest) are best characterized by low spatial resolution imagery (e.g., for monitoring trends in vegetation cover across North America). Conversely, high spatial resolution data are more appropriate for investigating disturbances that require a greater level of spatial detail, such as tree-level disturbances. For example, Figure 2.1 demonstrates that high spatial resolution data such as QuickBird or aerial photography would be required to capture tree-level damage caused by the mountain pine beetle.

The spatial extent of data sources must also be considered in conjunction with data costs. Low spatial resolution data sources typically cover larger spatial extents and are less expensive; therefore, the per unit cost for these data sources is less than medium- or high-resolution data sources. Conversely, high spatial resolution data sources generally have smaller spatial extents and higher per unit costs. In addition, high spatial resolution data also present additional challenges for project logistics; image files tend to be large and cumbersome to store, manipulate, and

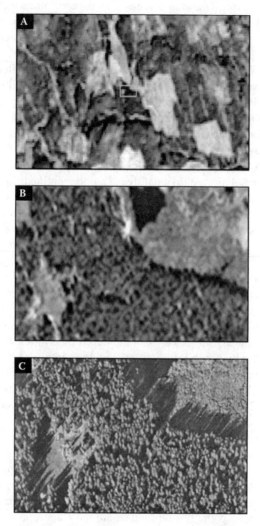

FIGURE 2.1 (See color insert following page 146.) Illustration of differing information content for three images with differing spatial resolution located near Merritt, British Columbia, Canada. Panel A is an approximately 8 km^2 area of 30-m spatial resolution Landsat 7 ETM+ multispectral imagery (Path 46/Row 25) collected on August 11, 2001. The 0.05-km^2 focus area in Panel A is represented in Panels B and C. Panel B is 2.4-m spatial resolution QuickBird multispectral imagery collected on July 17, 2004. Panel C is a digital ortho-image with a spatial resolution of 30 cm collected on August 22, 2003.

TABLE 2.1
Relationship Between Scale and Spatial Resolution in Satellite-Based Land Cover Mapping Programs

Spatial resolution	Nature of suitable forest disturbance targets
Low	Disturbances that occur over hundreds or thousands of meters (small scale); detectable with sensors such as GOES, NOAA AVHRR, MODIS, SPOT VEGETATION
Medium	Disturbances that occur over tens or hundreds of meters (medium scale); detectable with sensors such as Landsat, SPOT, IRS, JERS, ERS, Radarsat, and Shuttle platforms
High	Disturbances that occur over scales of centimeters to meters (large scale); detectable with aerial remote sensing platforms (e.g., photography), IKONOS, QuickBird

Source: Adapted from Franklin and Wulder, 2002.

process. Furthermore, the increased spectral variability of high spatial resolution imagery can confound many commonly used image classification methods (Wulder et al., 2004). Careful thought must therefore be given to the information need and the spatial resolution as higher spatial resolution data will not necessarily provide better information.

TEMPORAL RESOLUTION

The temporal resolution provides an indication of the time it takes for a sensor to return to the same location on the Earth's surface. The revisit time is a function of the satellite orbit, image footprint, and the capacity of the sensor to image off nadir. The timing of image acquisition should be linked to the target of interest. Some disturbance agents may have specific biowindows (e.g., fire, defoliating or phloem-feeding insects) during which imagery must be collected to capture the required information (Wulder, Dymond, et al., 2004), while other disturbances may be less specific (e.g., harvest). For ongoing programs designed to monitor forest change before and after a disturbance event, the acquisition of images should occur in the same season over a series of years (known as anniversary dates). Anniversary dates are critical to ensure that the spectral responses of the vegetation remain relatively consistent over successive years (Lunetta et al., 2004). In addition, a reduction in image quality may also occur due to nonoptimal sun angles and reduced illumination conditions; as a result, off-year imagery is typically preferred over off-season imagery for remote sensing mapping applications (Wulder, Franklin, et al. 2004). For some applications, however, the capacity to incorporate temporal resolution can be advantageous. For example, analysis of vegetation at both leaf-on and leaf-off times can provide important information on the pattern of understory vegetation and nondeciduous canopy condition (Dymond et al., 2002). Temporal resolution of airborne sensors is less critical as, in many cases, image collection is undertaken on demand, often coincident with insect outbreaks or fires (Stone et al., 2001).

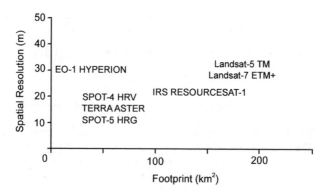

FIGURE 2.2 Comparison of spatial resolution and image footprint of current medium spatial resolution satellites (satellites and sensors listed in more detail in Table 2.2).

There are often trade-offs between image spatial and temporal resolution that have implications for data selection. Generally, high spatial resolution imagery has a smaller footprint (or image) size, and it takes longer for the satellite to revisit a location on the Earth's surface at nadir than broader scale imagery. However, many high-resolution sensors have the capacity to tilt or position the sensor at an angle, thereby allowing locations on adjacent swaths to be acquired. This results in satellites such as IKONOS and QuickBird (with short revisit times, varying from 1 to 3.5 days depending on latitude of target location); however, images acquired will be off nadir. Medium-resolution satellites such as Landsat revisit the same location once every 16 days. The relationship between spatial resolution and footprint (or image) size for other medium spatial resolution systems is presented in Figure 2.2.

SPECTRAL RESOLUTION

Spectral resolution provides an indication of the number and the width of the spectral wavelengths captured by a particular sensor The spectral resolution of standard black-and-white aerial photography is known as *panchromatic* and spans the complete visible portion of the electromagnetic spectrum, along with some portion of the near-infrared spectral wavelengths, with a single image band or channel. Sensors with more bands and narrower spectral widths are described as having an increased spectral resolution. Currently, most operational remote sensing systems have a small number of broad spectral channels: Landsat ETM+ data have seven spectral bands in the reflective portion of the electromagnetic spectrum and one band in the thermal-infrared region. Hyperspectral data (e.g., instruments with more than 200 narrow spectral bands) are becoming more widely available (Vane and Goetz, 1993), both on space-borne (such as the HYPERION sensor on the EO-1 [Earth Observer-1] platform) and airborne platforms such as HyMap (Cocks et al., 1998), *CASI* [Compact Airborne Spectographic Imager] (Anger et al., 1994), and the NASA AVIIS [Advanced Airborne Visible/Infrared Imaging Spectrometer] (Vane et al., 1993). The width and locations of these bands along the electromagnetic spectrum determine their suitability for forest disturbance applications. For example, a subtle spectral response, such as foliage discoloration, might manifest in a very specific region of the electromagnetic spectrum and may therefore be more effectively detected with a hyperspectral instrument, whereas a dramatic change, such as clearcutting, is discernible in a wide range of spectral wavelengths.

Remote sensing imagery is often categorized as either active or passive. Passive, or optical, remotely sensed data are collected by sensors sensitive to light in the 400- to 2500-nm region of the electromagnetic spectrum (encompassing the visible, near-infrared, shortwave, mid- and long-infrared regions of the spectrum), which includes detection of reflected light and temperature (such as weather or meteorological satellites). Passive remotely sensed data are the type most commonly used for vegetation studies and forest disturbance applications; Examples include aerial photography, Landsat, SPOT, IKONOS, and QuickBird.

Active remote sensing systems are characterized as those that emit energy, in one form or another, and then measure the return rate or amount of that energy back to the instrument. These active sensors can therefore operate under expanded meteorological conditions as the sun's illumination is not required. Microwave and LIDAR systems are examples of active sensors that provide the energy illuminating the surface and record the backscattered radiation from the target (Lefsky and Cohen, 2003). The most common implementation for microwave sensors is synthetic aperture radar, which utilizes microwave wavelengths of 1 mm to 1 m, about 2000 to 2 million times the wavelength of green light (500 nm) (Lefsky and Cohen, 2003). The choice of active versus passive systems for forest disturbance mapping will depend on the information need. Since active sensors can operate regardless of weather, they may be most effectively used in areas where there is perpetual cloud cover (e.g., tropical rain forests). Terrestrial LIDAR sensors typically capture a single spectral band, often between 900 and 1064 nm (Lefsky and Cohen, 2003). New disturbance mapping opportunities are enabled through the repeated collection of LIDAR data representing differing time periods, such as monitoring of forest gap dynamics and growth (St-Onge and Vepakomma, 2004).

RADIOMETRIC RESOLUTION

Radiometric resolution provides an indication of the actual information content of an image and is often interpreted as the number of intensity levels that a sensor can use to record a given signal (Lillesand and Kiefer, 2000). The finer the radiometric resolution of a sensor, the more sensitive it is to detecting small differences in reflected or emitted energy. Thus, if a sensor uses 8 bits to record data, there would be $2^8 = 256$ digital values available, ranging from 0 to 255. However, if only 4 bits were used, then only $2^4 = 16$ values ranging from 0 to 15 would be available, resulting in reduced radiometric resolution. Most low- and medium-resolution remotely sensed data commercially available are 8 bit. High-resolution data such as QuickBird are 11 bit. In terms of data selection for forest disturbance, radiometric resolution is the least critical of all of the image characteristics considered in this chapter as the sensors available for mapping of land cover and dynamics typically have a minimum of 8 bits. Given the option, it is usually better to use data with greater radiometric resolution; generally, users should receive data in the original bit format and not data that have been resampled to a lower radiometric resolution.

RESOLUTION INTERACTIONS AND IMPLICATIONS

The variety of remote sensors onboard the array of satellites operated by public and private agencies that are currently orbiting the Earth and collecting data at various spatial, temporal, radiometric, and spectral resolutions renders the compilation of an exhaustive list of remote sensing systems difficult. For a comprehensive listing of remote sensing instruments and missions, refer to the work of Glackin and Peltzer (1999) as well as to sources on the Internet, which provide additional details on existing and planned remote sensing systems (e.g., Stoney, 2004). Relevant attributes of the most common systems are summarized in Table 2.2, and an indication of commonly used sensors with a range of spatial and spectral characteristics is provided in Figure 2.3. When selecting an appropriate image source to capture forest disturbance information, the information needs of the end users must guide the selection of data with consideration of spatial, spectral, and temporal resolutions. Logistical issues, such as metadata, data storage, file manipulation, and data costs must also factor into the decision.

Once the relative merits of spatial, temporal, spectral, and radiometric properties have been considered relative to the target and information need, an appropriate data

TABLE 2.2
Characteristics of Select Low, Medium, and High Spatial
Resolution Sensors

Sensor	Footprint (km²)	Spatial Resolution (m) (*)	Spectral Resolution (nm)
Low Resolution Sensors			
NOAA 17 (AVHRR)	2940	1100	500–1250
SPOT 4 (VEGETATION)	2250	1000	430–1750
Terra (MODIS)	2330	500	366–14385
Medium Resolution Sensors			
Landsat-5 (TM)	185	30	450–2350
Landsat-7 (ETM+)	185	30 (MS/SWIR); 15 (pan)	450–2350
SPOT 2 (HRV)	60	20 (MS); 10 (pan)	500–890
SPOT 4 (HRVIR)	60	20	500–1750
SPOT 5 (HRG)	60	10 (MS); 20 (SWIR)	500–1730
IRS (RESOURCESAT-1)	141	23.5	520–1700
Terra (ASTER)	60	15	530–1165
EO-1 (HYPERION)	37	30	433–2350
High Resolution Sensors			
Orbview-3	8	4 (MS); 1 (pan)	450–900
QuickBird-2	16.5	2.44 (MS); 0.6 (pan)	450–900
IKONOS	13.8	4 (MS); 1 (pan)	450–850

* MS = multispectral, SWIR = shortwave infrared, pan = panchromatic.

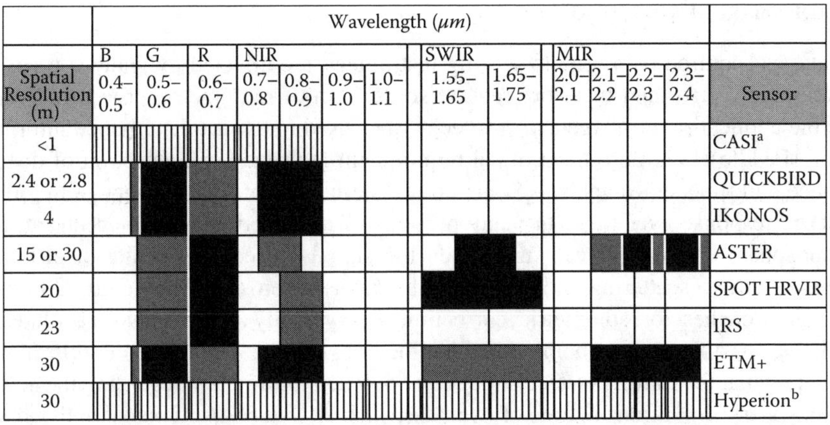

Spatial Resolution (m)	Wavelength (μm)													Sensor
	B	G	R	NIR				SWIR		MIR				
	0.4–0.5	0.5–0.6	0.6–0.7	0.7–0.8	0.8–0.9	0.9–1.0	1.0–1.1	1.55–1.65	1.65–1.75	2.0–2.1	2.1–2.2	2.2–2.3	2.3–2.4	
<1	▨	▨	▨	▨	▨									CASI[a]
2.4 or 2.8		■	■	■										QUICKBIRD
4	■	■	■	■										IKONOS
15 or 30		■	■		■				■	■	■	■	■	ASTER
20		■	■	■				■						SPOT HRVIR
23		■	■	■										IRS
30	■	■	■	■				■	■			■		ETM+
30	▨	▨	▨	▨	▨	▨	▨	▨	▨	▨	▨	▨	▨	Hyperion[b]

[a] CASI channels programmable in size; >2 nm width depending on application (Anger et al., 1994)
[b] Hyperion collects 220 bands of spectral data over the 400 to 2500 nm spectral range.

FIGURE 2.3 Spatial resolution and approximate spectral resolution of multispectral sensors commonly used for vegetation mapping. Shaded blocks represent different spectral bands. Blocks of narrower width tend to indicate a sensor with greater spectral sensitivity. [a] CASI channels programmable in size; > 2 nm width depending on application (Anger et al., 1994). [b] Hyperion collects 220 bands of spectral data over the spectral range of 400 to 2500 nm.

source may be selected. Following data selection, a series of preprocessing steps is typically required to prepare the data for further analysis. The preprocessing requirements are particularly necessary when multiple dates of imagery are used to characterize forest disturbance or change events in general. Radiometric and geometric processing methods are addressed in the following section.

RADIOMETRIC AND GEOMETRIC PROCESSING

Success in disturbance identification is dependent on robust radiometric and geometric preprocessing (Lu et al., 2004; Trietz and Rogan, 2004). Once the most appropriate remotely sensed imagery has been selected to monitor the disturbance and its spatial pattern, detection of this variation either spatially or temporally is only possible if changes in the phenomena of interest result in detectable changes in radiance, emittance, or backscatter (Smits and Annoni, 2000). Thus, it is critical that the change in signal is attributable to a real change in the land surface rather than a change in nonsurface factors such as atmospheric conditions, imaging and viewing conditions, or sensor degradation (Hame, 1988); radiometric processing is applied to image data to minimize the impacts of these factors on subsequent image analysis procedures. Similarly, the geometric matching of two or more scenes must be accurate, as image misregistration can have a large influence on the change detection results (Smits and Annoni, 2000). The following sections detail those processing steps required to prepare the imagery for further analysis.

RADIOMETRIC PROCESSING

Data, as acquired by a remote sensing instrument, are affected by many sources of radiometric error and therefore require some form of radiometric processing prior to the application of image analysis techniques used to extract disturbance information (Peddle et al., 2003). A critical requirement for successful detection of disturbance and time series analysis is the correct derivation of the true change in radiometric response over time. In many portions of the electromagnetic spectrum, the atmosphere has a significant impact on the signal sensed by satellite or airborne sensors due to scattering and absorption by gas and aerosols (Song et al., 2001).

Approaches to radiometric correction are typically described as absolute or relative (or a mixture of both). Absolute methods involve extracting the reflectance of a target at the Earth's surface and require detailed information regarding the actual atmospheric conditions at time of overpass, such as water vapor content and aerosol optical thickness, to adjust the imagery using radiative transfer theory (Peddle et al., 2003). A limitation to absolute atmospheric correction methods is the requirement for detailed atmospheric data that are rarely routinely available at the location or time of satellite overpass, especially when the analysis is retrospective. Relative radiometric correction methods are designed to reduce atmospheric effects and variability between multiple images by using common features in the two images that have invariant spectral properties (Chen et al., 2005). The choice of whether to use an absolute or relative radiometric correction method depends on many factors; refer to the work of Chen et al. (2005) and Song et al. (2001) for a more detailed discussion on the relative merits of each approach. It should be noted that some analysis methods have been developed using specific data types (e.g., ground surface reflectance); therefore, if the user intends to implement these methods, he or she must ensure that the data are corrected to the appropriate level. The topics included in this chapter cover fundamental radiometric considerations: conversion of raw image values or digital numbers (DNs) to radiance; conversion of radiance values to reflectance; and normalizing imagery to minimize the impact of different atmospheric or illumination conditions. A more thorough and detailed discussion of radiometric processing considerations is provided by Peddle et al. (2003) and Hall et al. (1991).

The methods described here, although generic in the sequence of steps that must be followed to complete the correction, are somewhat specific to Landsat products due to the long history of Landsat data usage. Research and methods for the radiometric processing of other sensors are becoming increasingly available (e.g., Pagnutti et al., 2003; Wu et al., 2005). Conversion of the sensor signal to surface reflectance requires that the raw DNs be first converted to radiance and then to reflectance. Conversion to at-satellite radiance (also known as top of atmosphere, TOA) is required if imagery from different sensors is to be compared (e.g., Landsat TM and ETM+) and is achieved using the following equation (Markham and Barker, 1986):

$$Rad_i = DN_i \times Gain_i + Offset_i \qquad (2.1)$$

where i is the band number for $i = 1, 2, 3, 4, 5, 7$; Rad_i is the TOA radiance of band i; DN_i is the DN of band i; $Gain_i$ is the gain of band i; and $Offset_i$ is the offset of band i.

Gains and offsets are provided in the header file for the imagery, or standard parameters specific to the sensor of interest are available from a variety of sources (e.g., Ekstrand, 1996; Huang et al., 2002; Markham and Barker, 1986).

The radiance values are then converted to reflectance using the following equation (Huang et al., 2002):

$$\text{Re } f_i = (Rad_i \times \pi \times d^2) / (ESUN_i \times \sin(\theta)) \tag{2.2}$$

where i is the band number for $i = 1, 2, 3, 4, 5, 7$; $\text{Re } f_i$ is the TOA reflectance of band i; Rad_i is the TOA radiance of band i; d is the Earth-Sun distance in astronomical units; $ESUN_i$ is the mean solar exoatmospheric irradiance of band i; and θ is the Sun elevation angle.

The Earth-Sun distance d can be determined by a lookup table based on the Julian day when the data were acquired (Irish, 2000). The mean solar exoatmospheric irradiances for Landsat-7 ETM+ bands are provided in Irish (2000), with information for other sensors also available (e.g., Pagnutti et al. 2003; Tuominen and Pekkarinen, 2005). The Sun elevation angle θ either can be found in the raw data header file or calculated based on the time and date of data acquisition. This conversion to TOA reflectance is necessary to correct for variation caused by solar illumination differences as well as cross-sensor differences in spectral bands.

When multiple images are used for change detection, disparities between the different image dates may persist (even after conversion to TOA reflectance) as a result of different atmospheric conditions and viewing and illumination geometries. To reduce these disparities, images undergo a normalization step (Du et al., 2002; Heo and Fitzhugh, 2000; McGovern et al. 2002; Yang and Lo, 2000). A number of variations on the normalization technique exist; however, most require use of a set of reference sites that appear over the entire image sequence. The sites, also known as pseudoinvariant features (Schott et al., 1988), are generally well-defined spatial objects in the scene that are interpreted as spectrally homogeneous and stable over time (Furby and Campbell, 2001). Both light and dark features can be used and often include lakes, mature even-age forest, dunes, and roads. Equations are then derived for all spectral channels to ensure these spectral features remain consistent over a temporal sequence of images (Yang and Lo, 2000).

GEOMETRIC CORRECTION AND IMAGE COREGISTRATION

In its raw state, satellite imagery contains spatial distortions that are a function of the acquisition system (e.g., factors associated with the sensor platform such as viewing angle, orbit, altitude, and velocity) or a function of external factors (e.g., effects of the Earth's curvature, relief displacement, and deformations resulting from map projections). Some of these distortions are systematic and are routinely corrected by the data vendor before the data are distributed. Other distortions are more difficult to fix and require the use of models or mathematical functions (Toutin, 2003). The term *geometric correction* refers to the processes used to correct spatial distortions; geometric correction is required to align remotely sensed imagery with other data sources and to combine multiple images, to either mosaic multiple images over large areas or coregister multiple images collected over the same location at different times. Geometric misregistration of images can be a significant source of error, and minimizing this error is a time-consuming task when undertaking change detection or data fusion methods (Dai and Khorram, 1998). Typically, a desirable target for geometric registration is an error of less than half a pixel. This ensures that misregistration does not introduce error into change detection results (Dai and Khorram, 1998; Igbokwe, 1999). It has been noted, however, that a misregistration, often reported as a root mean square error, of less than one pixel can be difficult to obtain (Gong and Xu, 2003).

Generally, all geometric correction methods require the collection of ground control points (GCPs), which are points concurrently identified from a corrected source (e.g., basemap, corrected image) and an uncorrected image source. The differences in the X and Y positions of these points between these two sources are used to compensate for spatial distortions in the uncorrected image. In the case of orthorectification, the Z position (or elevation) is also used for the correction. A summary of geometric correction methods are provided in Table 2.3, while a more detailed treatment of methods is provided by Toutin (2003, 2004). Geometric correction methods typically take one of two forms: parametric or nonparametric. Nonparametric methods are considered suitable for low-resolution imagery, while parametric methods are necessary for high-resolution imagery. In the context of mapping forest disturbance, geometric correction is critical if a change detection approach is used and if the resulting disturbance information is to be integrated into other spatial databases.

METHODOLOGIES FOR DISTURBANCE MAPPING

Once appropriate radiometric and geometric corrections have been applied, the image data are ready for analysis. The overriding objective when detecting landscape change and disturbances is to compare data from a series of points in time by (a) controlling all extrinsic factors caused by differences in variables that are not of interest and (b) assessing the real changes caused by the variable of interest (Lu et al., 2004). Therefore, as discussed in the previous section, minimizing and removing factors such as atmospheric attenuation and scattering, illumination, viewing distortion, and poor coregistration is critical to ensure the observed change is real. A wide variety of detection algorithms and time series approaches have been developed to detect change and disturbances in imagery, and selecting and implementing the most appropriate method is an important process in change detection studies. A number of current reviews exist (Coppin et al., 2004; Gong and Xu, 2003; Lu et al., 2004). Singh (1989) defined 11 categories of change detection techniques that can broadly be grouped into five distinct approaches: (a) image algebra (differencing, subtraction, or ratioing) of two or more images; (b) regression or correlation, in which a model is developed that predicts or compares spectral responses of a series of images; (c)

TABLE 2.3
Summary of Geometric Correction Methods

	Method	Description	Suitable applications/limitations
Nonparametric	2D polynomial functions	Methods commonly applied when a classic geometric correction is done Do not require any a priori information about the sensor and therefore do not reflect the source of distortions in the image First-order polynomials correct for translation in both axes, a rotation, scaling in both axes, and an obliquity Second-order polynomials additionally correct for torsion and convexity in both axes Third-order polynomial corrects for additional distortions, which do not necessarily correspond to any physical reality of the image acquisition system; third-order polynomial functions introduce errors in the relative pixel positioning in ortho-images	Limited to images with few or small distortions Most suitable for nadir viewing imagery, covering small areas, over flat terrain Not recommended when precise geometric positioning is required Not suitable for multisource/multiformat data integration and in high relief areas Requires numerous, regularly distributed GCPs Sensitive to error, not robust or consistent Correct locally at GCP locations only
	3D polynomial functions	An extension of 2D methods and the method typically used when a traditional orthorectification is complete Used when the parameters of the acquisition system are unknown	Limitations similar to 2D polynomials (see above) Most suitable for small images
	3D rational functions	Used to approximate a model previously determined with a rigorous 3D parametric function or to determine (via least-squares adjustment) the coefficients of the polynomial function	Have similar issues as 3D polynomial functions Should not be used with raw data or large-size images Use with small, georeferenced, or geocoded images Best choice among nonparametric methods
Parametric	3D parametric functions	Models the distortion of the platform, the Earth, and the cartographic projection	Depends on sensor, platform Most suitable method for high-resolution imagery

Source: Modified after Toutin, 2004.

statistical techniques such as the tasseled cap transformation (TCT) and principal component analysis (PCA) that computes statistical components, which are then compared for temporal changes; (d) classification comparisons, by which images are classified separately, and the resulting classifications are compared; and (e) the increasing use of tools that analyze images and other data sets within a geographical information system (GIS). Each of these methods is discussed in detail in the following sections.

IMAGE ALGEBRA

The use of simple algebraic operations to assess levels of change and disturbance through a time series of images is a commonly applied, relatively easy, and straight-forward technique. The approaches all have the common characteristic of selecting either constant or dynamic thresholds to determine through time when and if a change has occurred. In this category of methods, two aspects are critical for the change detection results: selecting suitable image bands or vegetation indices and selecting suitable thresholds to identify the changed areas (Lu et al., 2004). The most commonly applied index is the normalized difference vegetation index (NDVI), which is the normalized ratio of the near-infrared and red regions of the spectrum (Eq. 2.3).

$$NDVI = \frac{(NIR - R)}{(NIR + R)} \qquad (2.3)$$

where R is the reflectance in the red, and NIR is the reflectance in the near infrared.

In the near-infrared region of the spectrum, within-leaf scattering is high, and as a result, reflected radiation from the canopy is also high. Conversely, in the red component of the spectrum, high absorption by pigments results in low radiation reflection. Consequently, changes in vegetation amount and cover, as well as the photosynthetic capacity of the vegetation, are typically positively related to an increase in the difference between near-infrared and red radiation (Peterson and Running, 1989; Price and Bausch, 1995).

A number of additional indices are based on theory similar to NDVI, such as the enhanced vegetation index (EVI) (Eq. 2.4) and specialty indices that incorporate the shortwave and mid-infrared spectral regions (such as the NDVIc [NDVI fire index]) (Eq. 2.5) and the normalized burn ratio (NBR) (Clark and Bobbe, Chapter 5, this volume; Hudak et al., Chapter 8, this volume; Key and Benson, 2005) (Eq. 2.6).

$$EVI = G * \frac{NIR - R}{NIR + C_1 R - C_2 B + L} \qquad (2.4)$$

$$NDVIc = \frac{(NIR - R)}{(NIR + R)} * [1 - \frac{(SWIR - SWIR_{min})}{(SWIR_{max} - SWIR_{min})} \qquad (2.5)$$

$$NBR = \frac{(NIR - SWIR)}{(NIR + SWIR)} \qquad (2.6)$$

where B is the reflectance in the blue; R is the reflectance in the red; NIR is the reflectance in the near infrared; $SWIR$ is the reflectance in the shortwave or mid-infrared spectral channels; and G, C_1, C_2, and L are user-specified constants.

At the broad scale, Potter et al. (2003) utilized a sequence of long-term AVHRR monthly spectral vegetation indices from 1982 to 1999 to identify major global disturbance events. Monthly vegetation indices were compared to a derived 18-year long-term average. The majority of the disturbance events (predominantly fire related) occurred in tropical savannah, scrubland, or boreal forest ecosystems. The analysis concluded that nearly 9 Pg of carbon have been lost from the terrestrial ecosystem to the atmosphere as a result of large-scale ecosystem disturbances. At the landscape scale, Nelson (1983) used the difference of the near-infrared spectral channels from Landsat MSS to delineate areas of gypsy moth defoliation. Lyon et al. (1998) undertook a comparison of seven spectral indices from three different dates to detect land cover change and concluded that changes in NDVI provided the best detection of vegetation change. In addition to using Landsat data, imagery from other sensors can be incorporated. For example, Stow et al. (1990) found that ratioing red and near-infrared bands of a Landsat MSS–SPOT high-resolution visible image (HRV) multitemporal pairs produced substantially higher change detection accuracy (about 10% better) than ratioing similar bands of a Landsat MSS–Landsat TM multitemporal pair (Lu et al., 2004).

IMAGE REGRESSION OR CORRELATION

More advanced methods of change detection can include the use of geometric models, spectral mixture models, and biophysical parameter models. In these approaches, multidate change is computed from physically based parameters such as leaf area index or biomass values, which are in turn computed from reflectance values. These transformed variables are preferred over simple vegetation indices for facilitating the interpretation of change and the extraction of vegetation information (Hall et al., Chapter 4, this volume; Lu et al., 2004). Adams et al. (1995) applied spectral linear unmixing approaches to extract spectral end-members, including healthy vegetation, nonphotosynthetic vegetation, exposed soil, and shade, and then analyzed changes in these spectral members as surrogates for land cover change. Rogan et al. (2002) applied a similar approach using Landsat imagery. Within a biophysical model framework, combinations of spectral bands as well as other data such as climate can be used to assess disturbance and land cover change. For example, monitoring phenological patterns of vegetation and its subsequent change is possible using a range of techniques, including measures of similarity (Coops and Walker, 1996), Fourier analysis (Andres et al., 1994), wavelet theory (Meyer, 1990), and harmonic analysis (Jakubauskas et al., 2001).

Bennett (1979) provided a mathematical overview of spatial time series analysis. With these techniques, the emphasis is not only on temporal change but also on the shape characteristics of the temporal change. Lambin and Strahler (1994) used three indicators — vegetation indices, land surface temperature, and spatial structure derived from AVHRR — to detect land cover change. Lawrence and Ripple (1999) utilized eight Landsat TM scenes to monitor changes in vegetation through time using fitted statistical models between each date to assess changes in overall vegetation cover. A key advantage of using these profile-based techniques that link with other data sets such as climate is that the full variation in the phenological cycle is resolved as data are collected throughout the growing season. As a result, changes linked to seasonality can be separated from other land cover changes and disturbances. A disadvantage is that typically only coarse spatial resolution imagery has a high enough temporal frequency to develop the necessary temporal profiles. This limits the change categories that can be detected and monitored (Coppin et al., 2004), although some research has taken place using time series to monitor ecosystem disturbances at finer spatial resolutions (e.g., Coops et al., 1999; Rogan et al., 2002; Sawaya et al., 2003).

STATISTICAL TECHNIQUES

Rather than a simple ratio of spectral channels, more refined transformations of the input spectral bands have been promoted as a technique to extract information about vegetation disturbance. One advantage of statistical approaches is that they reduce data redundancy between bands and emphasize different information in derived components (Lu et al., 2004). The most commonly applied techniques are based on PCA and the TCT (Crist and Cicone, 1984). Although the use of principal components to derive multitemporal change can be difficult to ascertain without a detailed understanding of the eigen structure of the data, the link between vegetative change and TCT has been shown to be generally more robust (Collins and Woodcock, 1996; Coppin et al., 2004). Simplistically, the TCT are guided and scaled PCA, which transform the Landsat bands into channels of known characteristics; soil brightness, vegetation greenness, and soil/vegetation wetness. Changes in these components over time can therefore reflect changes in the vegetation characteristics.

Cohen et al. (1998) contrasted the brightness and greenness components of a TCT output to assess changes in forest biomass in the Pacific Northwest of the United States from 1976 to 1991 and found harvest activity was detected in over 90% of the known clearcuts. As the wetness component contrasts the sum of the visible and near-infrared bands with the longer infrared bands to estimate vegetation or soil moisture, it has been used with success to detect forest disturbances through time. The difference between wetness indices calculated for multiple dates (known as the enhanced wetness difference index) has been used to discriminate partial harvesting with a per-pixel accuracy of approximately 71% (Franklin et al., 2000).

This technique has also been applied by Skakun et al. (2003) to detect red-attack damage caused by mountain pine beetle (*Dentroctonus ponderosa* Hopkins) in stands of lodgepole pine (*Pinus contorta*). Skakun et al. (2003) used multitemporal Landsat ETM+ imagery that was corrected and processed using the TCT to obtain wetness components that were differenced to reveal spatial patterns of insect attack. Classification accuracy of red-attack damage based on this method ranged from 67% to 78%. In Figure 2.4, the use of TCT wetness to map mountain pine beetle red-attack

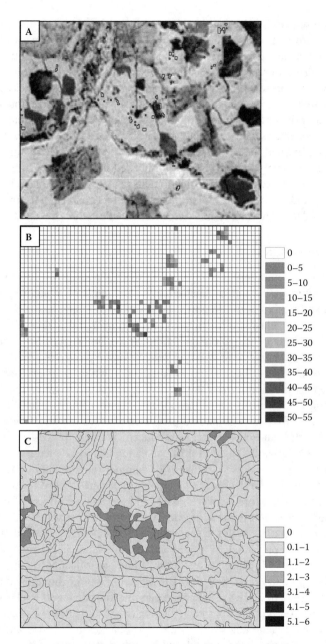

FIGURE 2.4 (See color insert following page 146.) Illustration of TCT wetness difference image with pixel-level insect infestation locations noted in yellow. Spatial information layers can be developed from the pixel-based infestation locations, such as Panel B, showing the pixel-based disturbance information aggregated as a proportion on a per hectare basis, and Panel C, in which the pixel-based disturbance is summed as an area estimate in hectares on a forest inventory polygon basis.

damage is presented. Pixel-based locations of insect attack are a single example of the types of information products that can be generated using this, or other, types of pixel-based change detection approaches. The Landsat pixel-based insect attack can be generalized to represent 1-Ha grid cells or forest inventory polygons. These grid or polygon representations of red-attack damage enable the pixel-based information to be ingested by models or to be incorporated into forest inventory databases.

Coppin and Bauer (1994) also examined changes in forest cover through use of the TCT components as well as simple vegetation indices (such as NDVI) and found that changes identified the most important forest canopy change features, and that these can be adequately expressed as a normalized difference. One key advantage of the TCT method over other statistical methods such as PCA, and as highlighted through these studies, is that the transformations are independent of the image scenes, while PCA is dependent on the image scenes (Lu et al., 2004).

IMAGE CLASSIFICATION

As an alternative to monitoring changes in the spectral response of vegetation before and after a disturbance event, another common technique of monitoring vegetation disturbance and pattern is to categorize all pixels in an image automatically into a series of land cover classes or themes and then compare the size and extent of the classes. This process of image classification can be either guided by human interpretation (known as *supervised classification*) or based principally on the statistical distribution of the spectral classes in the image (known as *unsupervised classification*). Image classification formed the basis of research investigating the differences in the structure and function of anthropogenic versus natural disturbance regimes (Tinker et al., 1998). Although natural processes (such as fire and windthrow) alter forest pattern, the landscape patterns produced by these processes is generally different from disturbances due to forest harvesting and associated road building. A single Landsat scene was used to classify a number of vegetation land cover and disturbance types. Several landscape pattern metrics were derived for the landscape as a whole and for the forest cover classes, and the relative effects of clearcutting and road building on the pattern of each watershed were examined. At both the landscape and cover class scales, clearcutting and road building resulted in increased fragmentation as represented by a distinct suite of landscape structural changes (Mladenoff et al., 1993; Tinker et al., 1998; White and Mladenoff, 1994).

A similar approach was adopted by Bresee et al. (2004), who utilized six images acquired from 1972 to 2001. A supervised classification was used to classify the six dominant land cover types in the area, including two disturbance classes, nonforested bare ground, and regenerating forest or shrub. Changes in the size and degree of fragmentation of each of the natural and disturbed land cover classes were then assessed over the 27-year period. Results indicated that changes in management objectives and natural disturbances have had a clear influence on landscape patterns and composition in the region throughout the past 30 years. The presence and temporal variability of windthrow events, disease outbreaks, and changes in stumpage value all greatly influenced the composition and structure of the forest stands (Bresee et al., 2004).

Cohen et al. (2002) compared over 50 Landsat scenes in the Pacific Northwest to monitor changes in disturbance patterns due to harvesting and fire over the past 30 years. An unsupervised classification approach was used to label pixels as disturbed, undisturbed, or confused. A trajectory for each pixel was then determined through time to provide overall maps of disturbance of the area. The historical imagery and mapping of spectral classes representing forest disturbance indicated that harvest rates were lowest in the early 1970s, peaked in the late 1980s, and then decreased again in the mid-1990s. By comparing managed and natural disturbance regimes through time, an understanding can be developed on the relative impact of management regimes on ecosystems (Cohen et al., 2002).

The comparison of two image classifications representing different dates to find change does need to be undertaken with care as the accuracy of each of the individual classifications effectively limit the accuracy of the final change layer (Fuller et al., 2003). For instance, if two classifications were to be used to find a 17% change with 75% reliability, both source classifications would require an accuracy of approximately 97% (Fuller et al., 2003).

GIS APPROACHES

The significant development of GIS and its widespread adoption in natural resource management, coupled with developments in modeling of terrain and climate, have resulted in the development and implementation of models that integrate remote sensing observations with other spatial data sets (Rogan and Miller, Chapter 5, this volume). The advantage of using GIS within a change detection analysis is the capacity to incorporate a range of data sources into each change detection application. Lo and Shipman (1990) used overlay techniques to detect urban development using multitemporal aerial photography and to map quantitatively, changes in land use. With the availability of different types of satellite imagery and the capacity to digitize and analyze maps, these GIS functions offer convenient tools for land use and land cover change detection studies (Lu et al., 2004), especially when the change detection involves long period or multiscale land cover change analysis (Petit and Lambin, 2001). This type of change detection, with its ability to combine multisource data sets, is the focus of ongoing research into the integration of GIS and remote sensing techniques for better implementation of change detection analyses.

OPERATIONAL CONSIDERATIONS

While the capability to monitor both vegetation disturbance and vegetation succession has been demonstrated with satellite and airborne image data sets (Foody et al., 1996), it is critical to recognize that disturbances not resulting in complete stand replacement (such as selective thinning) and successional processes that involve a slow change in species composition can be difficult to detect and classify (Table 2.4). Forest disturbance can be characterized by type (e.g., phenological, fire, disease, etc.); duration (e.g., days, months, years); spatial extent (e.g., tree, stand, watershed); rate (e.g., slow, medium, fast); and magnitude (e.g., small, medium, large) of the

TABLE 2.4
Major Types of Forest Change, Their Duration, Spatial Extent, Rate (on a Daily Basis), and Magnitude

Type of change	Time lapse (duration)	Spatial extent	Disturbance severity	Rate
Phenological	Days to months	All levels	Medium	Medium
Regeneration	Days to decades	Individual–stand	Small	Slow
Climatic adaptation	Years	All levels	Small	Slow
Wind throw/flooding	Minutes to hours	Individual–stand	Large	Medium to fast
Fire	Minutes to days	All levels	Large	Fast
Disease	Days to years	All levels	Small to large	Slow to medium
Insect attack	Days to years	All levels	Small to large	Slow to fast
Mortality	Days to years	All levels	Large	Slow to fast
Pollution	Years	Stand–watershed	Small to large	Slow
Silviculture (thinning/pruning)	Days	Stand–watershed	Large	Fast
Clearcutting	Days	Stand–watershed	Large	Fast
Plantation	Days to decades	Stand–watershed	Small	Fast

Source: After Gong, P. and Xu, B., in M.A. Wulder and S.E. Franklin (Eds.), *Remote Sensing of Forest Environments: Concepts and Case Studies* (pp. 301–333), Kluwer Academic Publishers, Norwell, MA, 2003.

disturbance. The interactions of these elements for a given disturbance combine to suggest the type of imagery that should be selected, the date range over which the images should span, and the area of coverage required.

The type of change (as identified in Table 2.4) has an influence on likelihood of detection using remote sensing image-based change detection procedures. Stand replacement disturbances (such as wildfire, clearcut logging) are more likely to be detected due to both their large visible extents and large change in vegetation structure and function (Cohen et al., 2002). In Figure 2.5, a relationship between the severity (or magnitude) and the accuracy that may be expected is portrayed. The notion is that subtle changes are more difficult to detect and map than dramatic changes. For instance, the removal of 10% of the stand volume to a partial harvest is more difficult to detect and map than a 40-Ha clearcut, resulting in lower mapped attribute accuracy or a lower detection likelihood. As a result, the expected accuracy when mapping changes in forest structure through partial harvesting is lower than when mapping clearcutting. The theory is supported through selected references included with Figure 2.5. The size (extent) of the disturbance also has an impact on the detectability as a function of the relationship between the spatial resolution and the objects of interest.

Following any classification or feature identification, some form of accuracy assessment is recommended, and requisite statistics for accuracy estimates should be calculated (Stehman and Czaplewski, 1998). It is important that independent training and validation data sets are used for the assessment of accuracy (Stehman,

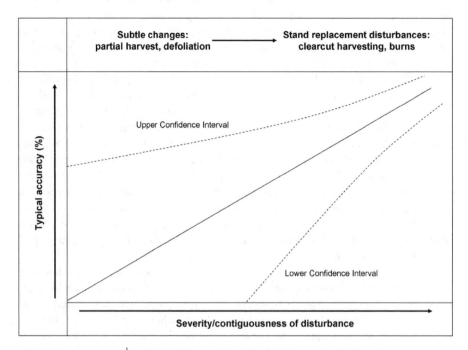

	Description	Accuracy Level	Reference
Non-Stand Replacement Disturbance	Defoliation	42–58%	Heikkila et al., 2002
	Partial cut	55–80%	Wilson and Sader, 2002
	Partial cut	55–70%	Franklin et al., 2000
	Partial cut	55–80%	Jin and Sader, 2005
Stand Replacing Disturbance	Wildire	74–98%	Wright Parmenter et al., 2003
	Clearcut, wildfire	88%	Cohen et al., 2002
	Wildfire	76%	Miller and Yool, 2002
	Clearcut	79–96%	Wilson and Sader, 2002
	Clearcut	>90%	Cohen et al., 1998

FIGURE 2.5 A theoretical representation of the increase in accuracy and decrease in confidence intervals (assuming equal sample sizes) associated with forest disturbance detection as disturbances on the forest landscape become more severe (e.g., increase in size) or more contiguous. Disturbances that are small and heterogeneous over the landscape, such as defoliation or partial harvesting, are generally more difficult to detect with remotely sensed data (depending on the spatial resolution of the data). Furthermore, the spectral variability associated with these disturbances is greater, making repeat detection of these non-stand-replacing disturbances less probable (i.e., the precision of these estimates is low). Conversely, larger and more spatially contiguous disturbances are generally mapped with greater consistency and greater accuracy, hence the narrowing of the confidence intervals for these stand-replacing disturbances.

1997). The data types that are commonly used are field and air photographs, other forms of purpose-collected data, and questioning or participation of knowledgeable stakeholders. The types of errors that emerge are characterized as either commission (falsely mapped changes) or omission (missed changes). The use of nonindependent data will typically yield a biased accuracy assessment (Rochon et al., 2003). Alternatively, if there is a lack of other independent observations with which to assess the accuracy of the output, statistical methods such as bootstrapping can help ensure that an unbiased estimate of the accuracy is developed. It is also acknowledged that the collection and use of training and validation data that reflect landscape changes can be problematic due to logistical and cost reasons. When mapping a single attribute of landscape disturbance, the collection of training and validation data are simplified by the number of classes under consideration; in this case, categorical transitions are from nondisturbed forest to some preidentified disturbance state, such as a harvest or insect attack. Analyses that are capturing a more broad range of changes require training and validation data that represent the full range of categorical transitions that are occurring or are expected to occur.

The accuracy assessment of the results of remote sensing change detection applications can be problematic due to the nature of the validation as it can be based on the process or the resultant products. The type, magnitude, and extent of the change (as presented in Table 2.4 and Figure 2.5) combine to influence the efficacy of the change detection approach. The nature of the change detection approach and the types of data used can also influence the ability of the analyst to capture the changes and the portrayal of the accuracy results. Operational limitations to validation are acknowledged, leading to an understanding that there is not a single best practice for the training and accuracy assessment of change detection results (Stehman et al., 2003).

CONCLUSIONS

In summary, when developing and applying remotely sensed time series data to assess forest change and disturbance, users should consider a range of important issues:

- Ensure the temporal and spatial scale of the disturbance phenomena monitored is well matched to the spatial, temporal, radiometric, and spectral resolution of the chosen remotely sensed imagery. In addition, ensure the data source can provide the information that the end user requires (e.g., a simple binary map showing disturbance areas versus a more complex product).
- Effective preprocessing is critical to successful forest disturbance detection and mapping. Once the imagery has been selected, it is crucial that the imagery is (or has been) calibrated to ensure that an observed change in signal is attributable to "true" change in the land surface rather than a change due to nonsurface factors such as different atmospheric conditions, imaging and viewing conditions, or sensor degradation. If multiple images are used (e.g., time series), then the images must be spatially aligned

precisely. High-quality geometric matching of the images is important to ensure that spurious change detection results do not occur.

- A variety of image-processing techniques exists to analyze change and detect disturbance regimes in remotely sensed observations. The method should be considered at the data selection stage, as not all data support all methods and vice versa. Select the most appropriate method (e.g., established or new spectral indices, statistical-based methods, image classification, or modeling) based on the desired outcome and level of complexity associated with the information needs of the end user.

- The increased use of GIS coupled with developments in modeling of terrain and climate have resulted in increasing interest in integrating changes in the spectral response with other spatial data sets within process-based modeling approaches. These models are providing useful information at regional and continental scales on ecological, hydrological, and physiological processes.

- Finally, some description or documentation of the accuracy of the disturbance or change mapping is required to provide users with an understanding of the reliability or limitations of the products produced. The description of the results of the change procedure can be heuristic or systematic and quantitative. The user can take the accuracy description and use this to guide the confidence placed on the change product for a given application.

REFERENCES

Adams, J. B., Sabol, D., Kapos, V., Filho, R.A., Roberts, D.A., Smith, M.O., Gillespie, and A.R. (1995). Classification of multispectral images based on fractions of endmembers: application to land-cover change in the Brazilian Amazon. *Remote Sensing of Environment, 52*, 137–154.

Andres, L., Salas W.A., and Skole, D. (1994). Fourier analysis of multi-temporal AVHRR data applied to land cover classification. *International Journal of Remote Sensing 15*, 1115–1121.

Anger, C., Mah, S., and Babey, S. (1994). Technological enhancements to the compact airborne spectrographic imager (CASI). *First International Airborne Remote Sensing Conference and Exhibition* (Vol. 2, pp. 205–213). Altarum Airborne Conferences, Ann Arbor, MI, U.S.A.

Bennett, R.J. (1979). *Spatial Time Series*. Pion, London, U.K. 674p.

Bresee, M., Le Moine, J., Mather, S., Crow, T., Brosofske, K., Chen, J., Crow, T.R., and Radmacher, J. (2004). Disturbance and landscape dynamics in the Chequamegon National Forest Wisconsin, USA, from 1972 to 2001. *Landscape Ecology, 19*, 291–309.

Chen, X., Vierling, L., and Deering, D. (2005). A simple and effective radiometric correction method to improve landscape change detection across sensors and across time. *Remote Sensing of Environment, 98*, 63–79.

Cocks, T., Jenssen, R., Stewart, A., Wilson, I., and Shields, T. (1998). The HyMap™ Airborne Hyperspectral Sensor: the system, calibration and performance. In M. Schaepman, D. Schläpfer, and K.I. Itten (Eds.), *Proceedings of the First EARSeL Workshop on Imaging Spectroscopy* (pp. 37–43). October 6–8, 1998, Zurich. EARSeL, Paris, France.

Cohen, W.B., Fiorella, M., Gray, J., Helmer, E., and Anderson., K. (1998). An efficient and accurate method for mapping forest clearcuts in the Pacific Northwest using Landsat imagery. *Photogrammetric Engineering and Remote Sensing, 64,* 293–300.

Cohen, W.B., and Goward, S.N. (2004). Landsat's role in ecological applications of remote sensing. *BioScience,* 54, 535–545.

Cohen, W.B., Spies, T.A., Alig, R.J., Oetter, D.R., Maiersperger T.K., and Fiorella, M. (2002). Characterizing 23 years (1972–1995) of stand replacement disturbance in western Oregon forests with Landsat imagery. *Ecosystems, 5,* 122–137.

Coops, N.C., Bi, H., Barnett, P., and Ryan, P. (1999) Prediction of mean and current volume increments of a eucalypt forest using historical Landsat MSS data. *Journal of Sustainable Forestry, 9,* 149–168.

Coops, N.C., and Walker, P.A. (1996). The use of the Gower metric statistic to compare temporal profiles from AVHRR data: a forestry and an agriculture application. *International Journal of Remote Sensing, 17,* 3531–3537.

Collins, J.B., and Woodcock, C.E. (1996). An assessment of several linear change detection techniques for mapping forest mortality using multitemporal Landsat TM data. *Remote Sensing of Environment, 56,* 66–77.

Colwell, R. N. (1960*). Manual of Photographic Interpretation.* American Society of Photogrammetry, Bethesda, MD. 868p.

Coppin, P., Jonckherre, I., Nackaerts, K., and Muys, B. (2004). Digital change detection methods in ecosystem monitoring: a review. *International Journal of Remote Sensing, 25,* 1565–1596.

Coppin, P.R. and Bauer, M.E. (1994). Processing of multitemporal Landsat TM imagery to optimize extraction of forest cover change features. *IEEE Geoscience and Remote Sensing, 60,* 287–298.

Crist, E.P. and Cicone, R.C. (1984). A physically-based transformation of Thematic Mapper data — the TM tasseled cap. *IEEE Transactions on Geoscience and Remote Sensing, 22,* 256–263.

Dai, X. and Khorram, S. (1998). The effects of image misregistration on the accuracy of remotely sensing change detection. *IEEE Transactions on Geoscience and Remote Sensing, 36,* 1566–1577.

Du, Y., Teillet, P.M., and Cihlar, J. 2002. Radiometric normalizations of multitemporal high-resolution satellite images with quality control for land cover change detection. *Remote Sensing of Environment, 82,* 123–134.

Dymond, C.C., Mladenoff, D.J., and Radeloff, V.C. (2002). Phenological differences in tasseled cap indices improve deciduous forest classification. *Remote Sensing of Environment, 80,* 460–472.

Ekstrand, S. (1996). Landsat TM-based forest damage assessment: correction for topographic effects. *Photogrammetric Engineering and Remote Sensing, 62,* 151–161.

Foody, G.M., Palubinskas, G., Lucas, R.M., Curran, P.J., and Honzak, M. (1996). Identifying terrestrial carbon sinks: classification of successional stages in regenerating tropical forest from Landsat TM data. *Remote Sensing of Environment, 55,* 205–216.

Franklin, S.E., Moskal, L.M., Lavigne M.B., and Pugh, K. (2000). Interpretation and classification of partially harvested forest stands in the Fundy model forest using multitemporal Landsat TM digital data. *Canadian Journal of Remote Sensing, 26,* 318–333.

Franklin, S.E. and Wulder, M.A. (2002). Remote sensing methods in medium spatial resolution satellite data land cover classification of large areas. *Progress in Physical Geography, 26,* 173–205.

Fuller, R.M., Smith, G.M., and Devereux, B.J. (2003). The characterisation and measurement of land cover change through remote sensing: problems in operational applications? *International Journal of Applied Earth Observation, 4,* 243–253.

Furby, S.L. and Campbell, N.A. (2001). Calibrating images from different dates to "like value" digital counts. *Remote Sensing of Environment, 77,* 1–11.

Glackin, D.L. and Peltzer, G.R. (1999). *Civil, Commercial, and International Remote Sensing Systems and Geoprocessing.* The Aerospace Press, El Segundo, CA. 89p.

Gong, P. and Xu, B. (2003). Remote sensing of forests over time: change types, methods, and opportunities. In M.A. Wulder and S.E. Franklin (Eds.), *Remote Sensing of Forest Environments: Concepts and Case Studies* (pp. 301–333). Kluwer Academic Publishers, Norwell, MA. 519p.

Goslee, S.C., Havstad, K.M., Peters, D.P., Rango, A. and Schlesinger, W.H. (2003). High-resolution images reveal rate and pattern of shrub encroachment over six decades in New Mexico, U.S.A. *Journal of Arid Environments, 54,* 755–767.

Goward, S.N. and Williams, D.L. (1997). Landsat and earth system science: development of terrestrial monitoring. *Photogrammetric Engineering and Remote Sensing, 63,* 887–900.

Hall, F., Strebel, D., Nickeson, J., and Goetz, S. (1991). Radiometric rectification: toward a common radiometric response among multidate, multisensor images. *Remote Sensing of Environment, 35,* 11–27.

Hame, T.H. (1988). Interpretation of forest changes from satellite scanner imagery. In *Satellite Imageries for Forest Inventory and Monitoring Experiences, Methods, Perspectives* (pp. 31–42). Department of Forest Mensuration and Management, University of Helsinki, Helsinki, Finland. Research Notes No. 21.

Heikkila, J., Nevalainen, S, and Tokola, T. (2002). Estimating defoliation in boreal coniferous forests by combining Landsat TM, aerial photographs and field data. *Forest Ecology and Management, 156,* 9–23.

Heo, J. and Fitzhugh, T.W. (2000). A standardized radiometric normalization method for change detection using remotely sensed imagery. *Photogrammetric Engineering and Remote Sensing, 66,* 173–182.

Huang, C., Wylie, B., Homer, C., Yang, L., and Zylstra, G. (2002). Derivation of a tasseled cap transformation based on Landsat 7 at-satellite reflectance. *International Journal of Remote Sensing, 23,* 1741–1748.

Igbokwe, J.I. (1999). Geometrical processing of multi-sensoral multi-temporal satellite images for change detection studies. *International Journal of Remote Sensing, 20,* 1141–1148.

Irish, R.R. (2000). Landsat-7 science data user's handbook. Retrieved from http://ltpwww. gsfc.nasa.gov/IAS/handbook/handbook_toc.html. National Aeronautics and Space Administration (last accessed April, 2006).

Jakubauskas, M.E., Legates, D.R., and Kastens, J.H. (2001). Harmonic analysis of time-series AVHRR NDVI data. *Photogrammetric Engineering and Remote Sensing, 67,* 461–471.

Jin, S. and Sader, S. (2005). Comparison of time series tasseled cap wetness and the normalized difference moisture index in detecting forest disturbances. *Remote Sensing of Environment, 94,* 364–372.

Key, C.H. and Benson, N.C. (2005). Landscape assessment: ground measure of severity, the composite burn index; and remote sensing of severity, the normalized burn ratio. In D.C. Lutes, R.E. Keane, J.F. Caratti, C.H. Key, N.C. Benson, and L.J. Gangi (Eds.), *FIREMON: Fire Effects Monitoring and Inventory System* (in press). USDA Forest Service, Fort Collins, CO, U.S.A.

Lambin, E.F. and Strahler, A.H. (1994). Indicators of land-cover change for change vector analysis in multitemporal space at coarse spatial scales. *International Journal of Remote Sensing, 15*, 2099–2119.

Lawrence, R.L. and Ripple, W.J. (1999). Calculating change curves for multitemporal satellite imagery: Mount St. Helens 1980–1995. *Remote Sensing of Environment, 67*, 309–319.

Lefsky, M.A. and Cohen, W.B. (2003). Selection of remotely sensed data. In M.A. Wulder and S.E. Franklin (Eds.), *Methods and Applications for Remote Sensing: Concepts and Case Studies* (pp. 113–46). Kluwer Academic Publishers, Norwell, MA. 519p.

Lillesand, T.M. and Kiefer, R.W. (2000). *Remote Sensing and Image Interpretation*. 4th ed. John Wiley and Sons, New York. 736p.

Lim, K., Treitz, P., Wulder, M., St-Onge, B., and Flood. M. (2003). LiDAR remote sensing of forest structure. *Progress in Physical Geography, 27*, 88–106.

Lo, C.P. and Shipman, R.L. (1990). A GIS approach to land-use change dynamics detection. *Photogrammetric Engineering and Remote Sensing, 56*, 1483–1491.

Lu, D., Mausel, P., Bronzdizio, E., and Moran, E. (2004). Change detection techniques. *International Journal of Remote Sensing, 25*, 2365–2407.

Lunetta, R.S., Johnson, D.M., Lyon, J.G., and Crotwell, J. (2004). Impacts of imagery temporal frequency on land-cover change detection monitoring. *Remote Sensing of Environment, 89*, 444–454.

Lyon, J.G., Yuan, D., Lunetta, R.S., and Elvidge, C.D. (1998). A change detection experiment using vegetation indices. *Photogrammetric Engineering and Remote Sensing, 64*, 143–150.

Markham, B.L. and Barker, J.L. (1986). Landsat MSS and TM post-calibration dynamic ranges, exoatmospheric reflectances and at-satellite temperatures. *EOSAT Landsat Technical Notes, 1*, 3–8.

McGovern, E.A., Holden, N.M., Ward, S.M., and Collins, J.F. (2002). The radiometric normalization of multitemporal Thematic Mapper imagery of the midlands of Ireland — a case study. *International Journal of Remote Sensing, 23*, 751–766.

Meyer, Y. (1990). *Ondelettes et opérateurs: I. Actualités Mathématiques*. Hermann, Paris. 215p.

Miller, J., and Yool, S. (2002). Mapping forest post-fire canopy consumption in several overstorey types using multi-temporal Landsat TM and ETM data. *Remote Sensing of Environment, 82*, 481–496.

Mladenoff, D.J., White, M.A., Pastor, J., and Crow, T.R. (1993). Comparing spatial pattern in unaltered old-growth and disturbed forest landscapes. *Ecological Applications, 3*, 294–306.

Nelson, R.F. (1983). Detecting forest canopy change due to insect activity using Landsat MSS. *Photogrammetric Engineering and Remote Sensing, 49*, 1303–1314.

Nelson, T., Wulder, M., and Niemann, K.O. (2001). Spatial resolution implications of digitising aerial photography for environmental applications. *Journal of Imaging Science, 49*, 223–232.

Pagnutti, M., Ryan, R.E., Kelly, M., Holekamp, M.,K., Zanoni, V., Thome, K., and Schiller, S. (2003). Radiometric characterization of IKONOS multispectral imagery. *Remote Sensing of Environment, 99*, 53–68.

Peddle, D.R., Teillet, P.M., and Wulder, M.A. (2003). Radiometric image processing. In M.A. Wulder and S.E. Franklin (Eds.), *Remote Sensing of Forest Environments: Concepts and Case Studies* (pp. 181–208). Kluwer Academic Publishers, Norwell, MA. 519p.

Perera, A.H. and Euler, D.L. (2000). Landscape ecology in forest management: An introduction. In A. H. Perera, D. L. Euler, and I. D. Thompson (Eds.), *Ecology of a Managed Landscape: Patterns and Processes of Forest Landscapes in Ontario* (pp. 3–11). University of British Columbia Press, Vancouver, BC, Canada. 346p.

Peterson, D.L. and Running, S.W. (1989). Applications in forest science and management. In G. Asrar (Ed.), *Theory and Application of Optical Remote Sensing* (pp. 429–473). John Wiley and Sons, New York. 734p.

Petit, C.C. and Lambin, E.F. (2001). Integration of multi-source remote sensing data for land cover change detection. *International Journal of Geographical Information Science, 15*, 785–803.

Potter, C., Tan, P.-N., Steinbach, M., Klooster, S., Kumar, V., Myneni, R., and Genovese, V. (2003). Major disturbance events in terrestrial ecosystems detected using global satellite data sets. *Global Change Biology, 9,* 1005–1021.

Price, J.C. and Bausch, W.C. (1995). Leaf area index estimation from visible and near-infrared reflectance data. *Remote Sensing of Environment, 52,* 55–75.

Rango, A., Goslee, S., Herrick, J, Chopping, M., Havstad, K., Huenneke, L., Gibbens, R, Beck, R and McNeely, R. (2002). Remote sensing documentation of historic rangeland remediation treatments in southern New Mexico. *Journal of Arid Environments, 50,* 549–572.

Richards, J.A. and Jia, X. (1999). *Remote Sensing Digital Image Analysis: An Introduction.* Springer, Berlin. 363p.

Rochon, G.L., Johannsen, C., Landgrebe, D., Engel, B., Harbor, J., Majumder S., and Biehl, L. (2003). Remote sensing as a tool for achieving and monitoring progress toward sustainability. *Clean Technologies and Environmental Policy, 5,* 310–316.

Rogan, J., Franklin, J., and Roberts, D.A. (2002). A comparison of methods for monitoring multitemporal vegetation change using Thematic Mapper imagery. *Remote Sensing of Environment, 80,* 143–156.

Running, S.W., Baldocchi, D., Turner, D., Gower, S.T., Bakwin, P., and Hibbard, K. (1999). A global terrestrial monitoring network integrating tower fluxes, flask sampling, ecosystem modeling and EOS satellite data. *Remote Sensing of Environment, 70,* 108–127.

Sawaya, K.E., Olmanson, L.G., Heinert, N.J., Brezonik, P.L., and Bauer, M.E. 2003. Extending satellite remote sensing to local scales: land and water resource monitoring using high-resolution imagery. *Remote Sensing of Environment, 30,* 144–156.

Schott, J., Salvaggio, C., and Volchok, W. (1988). Radiometric scene normalization using pseudoinvariant features. *Remote Sensing of Environment, 26,* 1–16.

Singh, A. (1989). Digital change detection techniques using remotely-sensed data. *International Journal of Remote Sensing 10,* 989–1003.

Skakun, R.S., Wulder, M.A., and Franklin, S.E. (2003). Sensitivity of the thematic mapper enhanced wetness difference index to detect mountain pine beetle red-attack damage. *Remote Sensing of Environment, 86,* 433–443.

Smits, P.C. and Annoni, A. (2000). Towards specification-driven change detection. *IEEE Transactions on Geoscience and Remote Sensing, 38,* 1484–1488.

Stehman, S.V. (1997). Selecting and interpreting measures of thematic classification accuracy. *Remote Sensing of Environment, 62,* 77–89.

Stehman, S.V. and Czaplewski, R.L. (1998). Design and analysis for thematic map accuracy assessment: fundamental principles. *Remote Sensing of Environment, 64,* 331–344.

Stehman, S.V., Sohl, T.L., and Loveland, T.R. (2003). Statistical sampling to characterize recent United States land-cover change. *Remote Sensing of Environment, 86,* 517–529.

Song, C., Woodcock, C.E., Seto, K.C., Lenney, M.P., and Macomber, S.A. (2001). Classification and change detection using Landsat TM data: when and how to correct atmospheric effects. *Remote Sensing of Environment, 75,* 230–244.

Spurr, S. (1948*). Aerial Photographs in Forestry*. Ronald Press Company, New York. 340p.

Stone, C., Kile, G., Old, K., and Coops, N.C. (2001). Forest health monitoring in Australia: national and regional commitments and operational realities. *Ecosystem Health, 7,* 48–58.

Stoney, W.E. (2004). *ASPRS Guide to Land Imaging Satellites*. Mitretek Systems. Accessed April 22, 2005 from http://www.asprs.org/news/satellites/satellites.html.

St-Onge, B. and Vepakomma, U. (2004). Assessing forest gap dynamics and growth using multi-temporal laser-scanner data. In M. Thies, B. Koch, H. Spiecker, and H. Weinacker (Eds.), *Proceedings of the ISPRS Working Group VIII/2. Laser-Scanners for Forest and Landscape Assessment* (Vol. 36, Part 8/W2, pp. 1682–1750). Freiburg, Germany, October 3–6, 2004. International Archives of Photogrammetry, Remote Sensing and Spatial Information Sciences. Available at http://www.isprs.org/commission 8/workshop_laser_forest/ST-ONGE.pdf.

Stow, D.A., Collins, D., and Mckinsey, D. (1990). Land use change detection based on multi-date imagery from different satellite sensor systems. *Geocarto International, 5,* 3–12.

Strahler, A.H., Woodcock, C.E. and Smith, J.A. (1986). On the nature of models in remote sensing. *Remote Sensing of Environment,. 20,* 121–139.

Thompson, M.M. and Gruner, H. (1980). Foundations in photogrammetry. In C.C.Slama, C. Theurer, and S.W. Henriksen (Eds.), *Manual of Photogrammetry* (pp. l–36). American Society of Photogrammetry, Falls Church, VA. 1056p.

Tinker, D.B., Resor, C.A.C., Beauvais, G.P., Kipfmueller, K.F., Fernandes, C.I., and Baker, W.L. (1998). Watershed analysis of forest fragmentation by clearcuts and roads in a Wyoming forest. *Landscape Ecology, 13,* 149–165.

Toutin, T. (2003). Geometric correction of remotely sensed images. In M.A. Wulder and S.E. Franklin (Eds.), *Remote Sensing of Forest Environments: Concepts and Case Studies* (pp. 143–180). Kluwer Academic Publishers, Norwell, MA. 519p.

Toutin, T. (2004). Geometric processing of remote sensing images: models, algorithms, and methods. *International Journal of Remote Sensing, 25,* 1893–1924.

Trietz, P. and Rogan, J. (2004). Remote sensing for mapping and monitoring land-cover and land-use change — an introduction. *Progress in Planning, 61,* 269–279.

Tuominen, S. and Pekkarinen, A. (2005). A. Local radiometric correction of digital aerial photographs for multi-source forest inventory. *Remote Sensing of Environment, 89,* 72–82.

Turner, M.G. (1989). Landscape ecology: the effect of pattern on process. *Annual Review of Ecology and Systematics 20,* 171–197.

Vane, G. and Goetz, A.F.H. (1993). Terrestrial imaging spectrometry: current status, future trends. *Remote Sensing of Environment, 44,* 117–126.

Vane, G., Green, R., Chrien, T., Enmark, H., Hansen, E., and Porter, W. (1993). The airborne visible/infrared imaging spectrometer (AVIRIS). *Remote Sensing of Environment, 44,* 127–143.

White, M.A. and Mladenoff, D.J. (1994). Old growth forest landscape transitions from pre-European settlement to present. *Landscape Ecology, 9,* 191–205.

Wilson, E.H. and Sader, S.A. (2002). Detection of forest harvest type using multiple dates of Landsat TM imagery. *Remote Sensing of Environment, 80,* 385–396.

Woodcock, C.E., Macomber, S.A., Pax-Lenney, M., and Cohen, W.B. (2001). Monitoring large areas for forest change using Landsat: generalization across space, time, and Landsat sensors. *Remote Sensing of Environment, 78,* 194–203.

Woodcock, C.E. and Strahler, A.H. (1987). The factor of scale in remote sensing. *Remote Sensing of Environment, 21*, 311–332.

Wright Parmenter, A., Hansen, A., Kennedy, R., Cohen, W., Langer, U., Lawrence, R., Maxwell, B., Gallant, A., and Aspinall, R. (2003). Land use and land cover change in the greater Yellowstone ecosystem: 1975–1995. *Ecological Applications, 13*, 687–703.

Wu, J., Wang, D., and Bauer, M.E. (2005). Image-based atmospheric correction of QuickBird imagery of Minnesota cropland. *Remote Sensing of Environment, 99*, 315–325.

Wulder, M., Hall, R., Coops, N., and Franklin, S. (2004). High spatial resolution remotely sensed data for ecosystem characterization. *BioScience, 54*, 1–11.

Wulder, M.A., Dymond, C.C., and Erickson, B. (2004). *Detection and Monitoring of the Mountain Pine Beetle*. Natural Resources Canada, Canadian Forest Service, Pacific Forestry Centre, Victoria, BC, Canada. Information Report BC-X-398. 28p.

Wulder, M.A., Franklin, S., and White, J. (2004). Sensitivity of hyperclustering and labeling land cover classes to Landsat image acquisition date. *International Journal of Remote Sensing, 25*, 5337–5344.

Yang, X, and Lo, C.P. (2000). Relative radiometric normalization performance for change detection from multi-date satellite images. *Photogrammetric Engineering and Remote Sensing, 66*, 967–981.

3 Remotely Sensed Data in the Mapping of Forest Harvest Patterns

Sean P. Healey, Warren B. Cohen,
Yang Zhiqiang, and Robert E. Kennedy

CONTENTS

INTRODUCTION

The economic value of the timber, paper, and fuel products extracted from forests has made forest harvesting one of the primary agents of forest disturbance globally. In addition to economic importance, forest harvests can have significant ecological consequences. Harvest levels have in some cases been observed to significantly affect wildlife habitat (Curtis and Taylor, 2004; Richards et al., 2002) and influence biogeochemical (Cohen et al., 1996; Hassett and Zak, 2005) and hydrological (Swank et al., 2001) cycles. This chapter summarizes a range of forest harvest practices by way of creating a framework for understanding data and method considerations specific to forest harvest detection. There are several thorough reviews of digital change detection methodology (Coppin et al., 2004; Gong and Xu, 2003; Singh, 1989). With those works as background, the focus here is on how the physical characteristics of various silvicultural operations influence data and method considerations in the change detection process. A case study involving detection of stand-replacing harvests in the Pacific Northwest of the United States of America is used to illustrate the process of selecting data and methods to address a particular set of analytical and mapping needs.

SILVICULTURAL OPTIONS

Silviculture is the science of manipulating ecological processes with a goal of shaping the structure of a forest stand to meet management goals. Harvesting, in addition to often yielding merchantable forest products, is a primary tool in this manipulation. The term *harvest* is used here to represent the cutting of any trees, even if that cutting does not produce salable products. In this section, harvests are broken into three different kinds of cuts: regeneration harvest, thinning, and salvage. This organization of harvest operations is based on silvicultural intent; regeneration harvest is intended to stimulate growth of a new cohort of trees, thinning is intended to stimulate growth of existing trees, and salvage is regeneration neutral, focusing mainly on extracting dead or dying trees that would otherwise become unusable. Salvage harvests have also on occasion been classified as a type of thinning when undertaken to remove suppressed trees proactively (e.g., Smith et al., 1997). However, only salvage that follows another disturbance is discussed in this chapter.

Although it is true that most remotely sensed data are fundamentally quantitative and not always suited to identifying silvicultural distinctions based on long-term intentions, there are nevertheless good reasons for becoming familiar with silvicultural theory. First, the management records often used as reference data in harvest mapping projects are likely to describe harvests in silvicultural terms (e.g., Franklin et al., 2000). Also, awareness of typical harvest strategies, some of which include a number of stand entries, can facilitate interpretation of the spectral data. This is particularly true if certain assumptions based on forest type and ownership can be used to predict likely harvest strategies in a given region. This section describes differences between regeneration, thinning, and salvage cuts and discusses the role of these harvest types within specific silvicultural systems.

REGENERATION HARVEST

Regeneration harvest (often called stand-replacing harvest) involves the removal of trees with the intention of stimulating the development of a new cohort of trees. In silvicultural systems using even-age stands, regeneration harvests can take the form of a clearcut, where all trees are removed, or a retention harvest, where some trees are allowed to survive for a variable length of time into the next rotation (Figure 3.1). Clearcuts are often chosen by industrial forest landowners because they are simple to administer and because they create exposed conditions that favor a large group of commercially important shade-intolerant species (Oliver, 1981). There are several reasons that a landowner may temporarily retain trees in what otherwise

Clearcut Regeneration Harvest

Seed Tree Retention Regeneration Harvest

Mechanical Thinning

FIGURE 3.1 (See color insert following page 146.) Common harvest practices represented in color orthophotos and tasseled cap-transformed (Crist and Cicone, 1984) Landsat data. Both sets of images were acquired in 2002 and show closed-canopy coniferous forests in central Washington State.

would be a clearcut: as a seed source, to moderate microclimatic conditions for the new cohort in a "shelterwood" capacity, or simply to diversify the structure of the succeeding stand for wildlife or aesthetic purposes (British Columbia Ministry of Forests, 2003). Trees in these systems may be left in densities and spatial distributions that are as various as the reasons for leaving them. The confusing aspect of retention cuts from the point of view of the remote sensing analyst is that even if retained trees are left in detectable densities, their sometimes gradual removal can occur at a stage when the rest of the stand is growing vigorously.

Related to even-age regeneration harvests are conversion cuts. Cutting all or most of the trees in a stand for the purpose of changing to a nonforest land use is not technically a silvicultural treatment, so it has no place in established silvicultural systems. Nevertheless, forest conversion is common throughout the world, particularly in the tropics, where shifting agricultural systems are employed (Achard et al., 2002). Losses of forest are also common on the forest/urban interface (Kline et al., 2004). While initially indistinguishable from even-age regeneration harvests, conversion cuts become identifiable over time as they fail to show a spectral signature consistent with forest recovery.

Some regeneration harvests are designed to create a multiage stand structure. Often called *selective harvests*, these cuts take out at least enough canopy and understory trees to stimulate regeneration of a new cohort. The intensity of removal can vary depending on the growth properties of the new trees, the preharvest structure of the stand, and both long- and short-term economic concerns. The management of multiage forests is considerably more complex than even-age forest management, primarily because competition must be managed between cohorts as well as within cohorts. Commitment to this more complex management by industrial forest owners is not common (Maguire, 2005); in practice, true selection harvests typically occur in places where, perhaps for wildlife or aesthetic reasons, uneven stand structure is a goal in itself.

THINNING

The silvicultural distinction between partial regeneration harvest and thinning is that regeneration harvest is intended to stimulate the growth of a new cohort of trees, whereas thinning is not. Smith et al. (1997) outlined four distinct approaches to thinning that are summarized here: low thinning, crown thinning, selective thinning, and geometric thinning. Low thinning involves the removal of trees below a chosen size. Usually, this means removing trees from suppressed and intermediate crown classes to benefit trees in co-dominant and dominant positions. This system is traditionally most common when suppressed trees are merchantable, often as firewood (Smith et al., 1997). A considerable amount of attention has recently been given to this type of thinning as a means to reduce forest fuels in fire-prone landscapes in the American west (Brown et al., 2004; Fight, 2004), although other types of thinning have also been considered (Fiedler, 2004). Low thinning has likewise been considered as a way to remove the less-vigorous trees that are particularly vulnerable to insect attack (Oliver and Larson, 1996). Recent forest legislation makes thinning that reduces forest health risks, a management priority (Healthy Forests Restoration Act, 2003), so low thinning may occur on public lands with increasing frequency.

As opposed to low thinning, which eliminates trees with growth rates that have already been suppressed, crown thinning involves the harvest of trees in the upper canopy classes to favor more vigorous or better-formed competitors. This practice is much more common than low thinning on industrial forest land for at least two reasons. First, the products of crown thinning are more often merchantable than the by-products of low thinning. Second, since it is large, vigorous trees that are removed, more of the site's resources are released to remaining trees, resulting in larger gains in the annual increment of the "crop" trees (Smith et al., 1997). Thinning of dense young stands also falls into this category, although the trees cut in these cases typically have limited commercial value. Such density management thins are common not only in intensively managed plantations, but also in naturally regenerated stands (Doruska and Nolen, 1999; Wilson and Oliver, 2000). Smith et al. (1997) suggested that this type of thinning makes most sense on poor sites where the limitation of a critical resource prevents structural differentiation of the canopy into distinct strata.

The third type of thinning, selective thinning, may appear quite similar to crown thinning in that both involve the thinning only of trees in the upper strata. The difference between the two approaches lies in the selection of trees to be left. Crown thinning involves cutting co-dominants to favor dominant trees, whereas in selective thinning dominants are cut to favor promising trees in co-dominant or subordinate crown positions. A practice that may resemble either harvest type involves the removal of trees with more regard for species than canopy class. This practice is used in the tropics in places where only a few trees per hectare may be merchantable.

Geometric thinning is the removal or destruction of all trees in a regular pattern, usually in lines or nonlinear strips (Figure 3.1). This type of thinning is most often used as an expeditious way to reduce density in intensively managed young plantations, although it is also used in older, less-intensively managed stands where growth has stagnated and canopy differentiation has not occurred (Smith et al., 1997). In practice, it is common for more than one of the above thinning systems to be combined because of multiple management goals or the relatively high cost of each entry. Thus, the possibility of significant removals from all canopy classes should be allowed in the measurement of thinning activities.

SALVAGE

Salvage cuts are used to extract wood products from stands that have been injured by insects, fire, disease, or some other disturbance. Salvage harvests are often aggressively pursued by industrial forest landowners because of their need to recover the economic value of the dead trees. However, salvage is less common by public agencies, which are often charged to manage forests for both economic and non-economic values. Accessibility also influences the likelihood of salvage cuts; trees have a finite window of merchantability after mortality, and there often is not enough time to extend the infrastructure necessary for salvage.

Salvage cuts are different from the other types of harvest in that the harvested trees' photosynthetic contribution to canopy reflectance has either ended or been diminished prior to removal. This makes it difficult to detect a salvage thinning as an event separate from the initial disturbance. While new forest roads in a stand following a disturbance may be an indication of harvesting, many salvage operations are conducted by helicopter (Donovan, 2004), and others use existing road networks. Because of the fundamental difficulty in separating salvage cuts from the mortality resulting from the disturbances that precede them, salvage cuts are not considered in the following sections.

DATA CONSIDERATIONS FOR HARVEST DETECTION

Coops et al. (Chapter 2, this volume) describe major issues involved with using remotely sensed data to detect forest disturbances. With Chapter 2 as background, this section focuses on considerations of remotely sensed data that are specific to the detection and characterization of harvests. As described in the preceding section, harvest types vary in tangible ways, including: the stratum of the canopy that is removed, the vigor of remaining trees, the expected response of the understory, and the patchiness of removal. Since all of these factors can influence the reflective properties of a stand, this section makes use of previously described silvicultural distinctions as a framework for understanding the spatial, temporal, and spectral resolution issues surrounding the remote detection of harvest.

SPATIAL PROPERTIES

The spatial resolution required of the data supporting harvest mapping depends largely on the dimensions of the forest changes to be mapped. Woodcock and Strahler (1987) suggested that spatial resolution should approximately match the spatial scale of the features of interest. If cells are much larger than the features to be detected, the spectral signal of the features will be averaged with and perhaps lost in the "noise" of spectrally dissimilar neighboring areas (Lefsky and Cohen, 2003). If spatial resolution is significantly finer than features of interest, cell values will display the reflective properties not of the features themselves but of their physical components, resulting in unnecessary within-feature spectral variation.

The most commonly used remotely sensed data for harvest mapping have come from sensors such as Landsat and SPOT (Système pour l'Observation de la Terre) that have functional resolutions of 10–100 m (e.g., Alves et al., 1999; Cohen et al., 2002; Franklin et al., 2000; Heikkonen and Varjö, 2004; Jin and Sader, 2005; Sader et al., 2003; Saksa et al., 2003; Varjö, 1996; Woodcock et al., 2001). SAR (synthetic aperture radar) data have also been resampled to this resolution for harvest mapping (Ranson et al., 2003). This grain size is fine enough in relation to the size of most harvest units that the ratio of boundary pixels influenced by exterior conditions is relatively small compared to the number of interior pixels. However, this resolution does not allow identification of individual or small groups of trees that have been either cut or retained through the harvest process. While the presence of these trees may be inferred through modeling (e.g., Collins and Woodcock, 1996; Healey et al., 2006), the spatial averaging that occurs in the Landsat/SPOT range of resolution precludes explicit mapping of the removal and retention patterns that distinguish the various harvest practices described in this chapter.

A variety of sensors, including digital frame camera platforms such as Airborne Data Acquisition and Registration (ADAR) and satellite-based systems like IKONOS and QuickBird have the ability to detect removal of large individual trees (Clark et al., 2004; Read et al., 2003). Although high-resolution sensors have been little used in regional harvest detection, they represent a potential means to better distinguish cuts that target co-dominant trees (i.e., crown thinning and selective harvest) from those that target typically broader dominant trees (i.e., selective thinning and high-grading) and those that remove no canopy trees (low thinning). The low thinning that is obscured from passive optical systems by the overstory could potentially be directly measured with canopy-penetrating technologies such as interferometric synthetic aperture radar and laser altimetry (LIDAR). However, to our knowledge, no work on this approach has been published, however.

Coarser spatial resolution data, including data from the AVHRR (advanced very high resolution radiometer) and MODIS (moderate resolution imaging spectroradiometer) platforms, have been used to map land cover changes across large areas

(Lambin and Ehrlich, 1997), but individual disturbances can be difficult to isolate at that scale (Ehrlich et al., 1997). Zhan et al. (2002), using 250-m MODIS data, were able to detect large-scale natural disturbances such as fire and flooding with relatively high accuracy. They found mapping of small-scale harvests, however, to be more problematic. Potter et al. (2005) used time series of 8-km AVHRR data to identify sudden decreases in fraction of photosynthetically active radiation over 20 years in the Pacific Northwest. According to finer-scale maps produced with Landsat data, these events roughly corresponded to large wildfires and periods of accelerated harvests. For calibrating global carbon models, the purpose and scale for which it was designed, this algorithm produced satisfactorily detailed information about harvest levels.

For mapping individual harvests, however, the majority of harvest detection projects have used medium spatial resolution data (e.g., Landsat). Although the silvicultural distinctions that may be possible through the use of high spatial resolution imagery (e.g., IKONOS) would communicate a good deal about managerial intentions and the likely future stand development, the tradeoff discussed in Chapter 2 of this volume between spatial resolution and spatial extent, has thus far limited the role of such data in harvest detection.

Temporal Properties

The element of change in harvest mapping places particular emphasis on the temporal dimensions of the source data set. The remeasurement frequency must match the duration of the harvest signal, and measurements must be available that cover the time period of interest. Before discussing issues of temporal resolution, the relatively straightforward issue of temporal extent is mentioned. If digital source data is required for a period prior to the mid-1990s, the choice of platforms includes CORONA (operational from 1959 to 1972), Landsat (launched in 1972), AVHRR (launched in 1978), and SPOT (launched in 1987). This is not to suggest, however, that harvest detection must limit itself to historical data sources as it moves forward. For example, several harvest-mapping programs have successfully bridged the spectral and spatial differences between Landsat Multi-Spectral Scanner (MSS) and Landsat Thematic Mapper (TM imagery) (Alves et al., 1999; Cohen et al., 2002; Lunetta et al., 1998; Woodcock et al., 2001). Rigina (2003) combined declassified CORONA panchromatic high-resolution imagery from 1964 with both Landsat and Indian Remote Sensing 1C panchromatic imagery from 1996 to study forest decline and disturbance in Russia. The same objectives could also be accomplished with digitized historical aerial photos. Further cross-platform efforts should be expected as the constellation of remote sensing platforms evolves.

For the sake of simplicity, comparative discussions of temporal resolution in this section assume a Landsat-like data source of medium spatial and spectral resolution. A distinction must be made between the temporal resolution of a particular sensor (i.e., potential overpass frequency) and the temporal resolution of the data set assembled for a harvest detection project. While it may be beneficial to possess a temporally dense sample of remotely sensed data, particularly to the extent that it allows separation of harvest and phenological change (Coppin et al., 2004), the cost of acquiring and coregistering imagery can strictly limit the number of dates considered (e.g., Lunetta et al., 1998). There are also cases in which very high temporal resolution can be counterproductive. If a harvest is carried out over months or years as a series of light removals, the spectral signal of what in aggregate may be a significant harvest could be rendered imperceptible when spread out over several monitoring periods. This scenario is most likely in areas where shelterwood retention harvests are common.

More common, though, is the concern that if the monitoring interval is too long, harvest signal will be lost or diluted because of postharvest canopy regrowth. This regrowth can take the form of regeneration of desired species and their competitors in the case of regeneration harvests, or it can take the form of canopy expansion in trees remaining after a thinning. Regeneration harvests generally produce a spectral signal that may be long-lasting (i.e., several years) but is also dynamic. Although the trees that regenerate following these harvests often grow quickly because of newly abundant site resources, the time required for this regeneration to appear and then the rate at which it grows can be highly variable across a landscape (Yang, 2004). Ecosystem productivity may play a role in how quickly harvest signal decays (Healey et al., 2005) and so may ownership-dependent variables such as pre- and postharvest site preparation. A survey of some of the many projects focused on detection of clearcuts (Banner and Ahern, 1995; Cohen et al., 2002; Hayes and Sader, 2001; Masek, 2005; Saksa et al., 2003) revealed remeasurement intervals ranging between 2 and 14 years. As a general rule, if harvest is to be mapped in terms of intensity and not just occurrence, fairly high temporal resolution will be required to minimize the degenerative effects of growth on the harvest signal.

Although no study based solely on remotely sensed data has characterized harvests according to the silvicultural framework outlined, a few have partial harvests as a separate class. Franklin et al. (2000), working in New Brunswick, and Fischer and Levien (2001) in California were able to map harvests with fairly high accuracies into broad categories of intensity using Landsat TM data separated by five years. Working in Maine, Jin and Sader (2005) found that while five-year monitoring intervals produced clearcut maps nearly as accurate as those using two-year intervals, annual or biennial imagery was preferable for characterizing less-intensive harvests. Healey et al. (2006) used two-year intervals in Washington to produce continuous estimates of cover loss and basal area removal. In choosing the remeasurement frequency for a harvest-mapping project, consideration should be given to the stability of the harvest signal in relation to the level of detail required by the change detection process (recall Figure 2.5). Signal decay is influenced both by regional harvest practices and local revegetation rates. If fine distinctions between harvest levels are to be mapped or if light removals must be detected, then more frequent data collection will be needed.

Spectral Properties

For the detection of harvests, the ideal remotely sensed data will sample the areas of the electromagnetic spectrum that best separate the forest canopy from whatever mix of soil and vegetation is exposed when trees are removed. The better the contrast

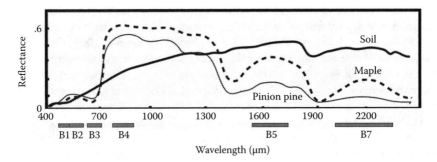

FIGURE 3.2 Reflectance patterns of three physical components of a study scene. (Adapted from Lefsky, M.A. and Cohen, W.B., in M. Wulder and S. Franklin, Eds., *Methods and Applications for Remote Sensing: Concepts and Case Studies* (pp. 13–46), Kluwer Academic Publishers, Norwell, MA, 2003, with kind permission of Springer Science and Business Media.) Wavelengths measured by the Landsat TM bands (B1-B5, B7) are also shown.

in reflectance between disturbed and undisturbed forest, the more easily canopy removal can be measured. Factors that influence the spectral contrast between canopy and background include species composition, soil properties, shadow patterns, and canopy density. The ability to predict wavelengths at which reflectance from canopy and background are most different can help not only in the choice of satellite, but also in targeting useful bands.

Figure 3.2 presents end member reflectances of soil, conifer needles, and deciduous leaves from a local spectral library (modified from Lefsky and Cohen, 2003). For simplicity, if these three elements were the only reflective components of a scene, then the general response in any spectral region could be predicted for a given harvest practice. Since the three components in Figure 3.2 differ little at the 400- to 530-nm range, a relatively slight response would be expected if a harvest was to realign their relative weights in the composite spectral response. Alternately, a substantial signal would be observed in the 1600-nm region if a harvest removed the conifer component, leaving an equal mix of hardwoods and soil reflectance. If the background were made up purely of soil, then the harvest signal at this range would be even more pronounced.

Familiarity with the spectral effects of the harvest practices to be encountered can inform the choice of the spectral characteristics of a project's source data. The above scenario is, of course, a simplification of the factors affecting reflectance. Factors like logging slash, skid trails, herbaceous composition, species shifts, and shadow can make the interpretation process more complex (Danson and Curran, 1993; Franklin et al., 2000; Nilson et al., 2001; Olsson, 1994). A number of harvest mapping projects have focused on the contrast between reflectance in the visible range (particularly in the red), which is strongly absorbed by vegetation, and the near infrared, which is not (e.g., Banner and Lynham, 1981; Hayes and Sader, 2001; Miller et al., 1978; Singh, 1989; Zhan et al., 2002). The majority of these studies utilized transformations such as the normalized difference vegetation index to accentuate the red/near-infrared contrast.

Strong evidence has emerged that the shortwave infrared (SWIR) portion of the spectrum contains information that greatly improves characterization of changes in forest structure (Skole and Tucker, 1993; Cohen and Goward, 2004; Collins and Woodcock, 1996; Lu et al., 2004; Skakun et al., 2003; Williams and Nelson, 1986). As can be seen in Figure 3.2, significant separation exists between soil and vegetation reflectance in the SWIR range (B5 and B7), so reducing the fractional contribution of vegetation through harvest should result in higher reflectance in this range. This is generally the case, while visible reflectance typically also increases, and near-infrared reflectance decreases (Franklin et al., 2000). Figure 3.3a illustrates this pattern as observed in a study of partial harvest in Washington (Healey et al., 2006). The similarity between these data and results from Finland presented by Olsson (1994) suggest a relatively consistent response among regions. The wetness index of the tasseled cap transformation (Crist and Cicone, 1984), which is a contrast between the sum of the visible/near-infrared bands and the sum of the SWIR bands, is commonly used with Landsat data to amplify the changes in SWIR in harvest detection (Cohen and Goward, 2004; Franklin et al., 2000). Healey et al. (2006) found a strong relationship between wetness change and degree of basal area removal (Figure 3.3b).

Although the wetness transformation is orthogonal to the other tasseled cap indices (Crist and Cicone, 1984), the results presented in Figure 3.3b clearly suggest a strong negative correlation between brightness and wetness in forest change space. This putative correlation was used by Healey et al. (2005) to reduce tasseled cap space to a single axis called the disturbance index, which is oriented in the presumed direction of increasing canopy removal. Jin and Sader (2005) mapped harvests using the normalized difference moisture index (NDMI), which contrasts near infrared and SWIR. NDMI-derived maps were similar in accuracy to maps produced with tasseled cap wetness, and NDMI can be used with sensors for which no tasseled cap transformation has been developed.

The above studies were carried out with sensors of low-to-medium spectral resolution. Little use has thus far been made of hyperspectral imagery to map harvests. Likewise, although a number of groups are currently investigating the potential for using LIDAR in harvest detection, little has been published in this area (Lefsky and Cohen, 2003). A good deal of work has been done with SAR in the microwave portion of the spectrum to detect clearcuts and conversion cuts. Intensity of backscattering in this range can be significantly different in forests and clearings, and this difference can be used to map canopy openings (Drieman, 1994; Ranson et al., 2003; Saatchi et al., 1997; Yatabe and Leckie, 1995). Coherence of the return signal phase between two dates can also be used to identify harvests (Antikidis et al., 1998; Smith and Askne, 1997, 2001; Wegmüller and Werner, 1995), although care must be taken in image acquisition because radar imagery can be sensitive to seasonal (Cihlar et al., 1992) and weather differences (Smith and Askne, 2001). Texture images have also been derived from SAR data and have been used to aid in the identification of clearcuts (Luckman et al., 1998; Sanden and Hoekman, 1999). Measurements of partial harvests have not yet been made with SAR, however with different wavelengths of microwave energy able to penetrate to different depths of the canopy (Jet Propulsion Laboratory, 1986; Treuhaft et al., 2004), there exists the

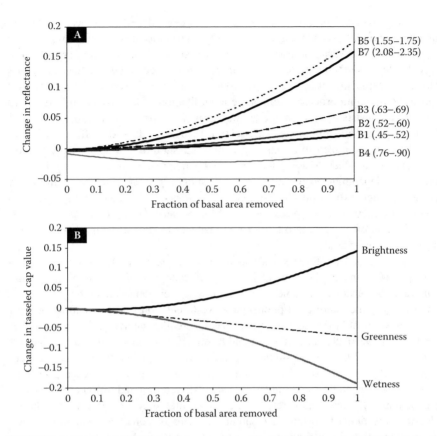

FIGURE 3.3 Changes in Landsat reflectance values (a) and tasseled cap indices (b) associated with a range of basal area removals in central Washington. Basal area removal was inferred from stump data combined with harvest permit records. Digital counts of Landsat imagery corresponding to pre- and postharvest conditions were converted to ground reflectance units using the COST model described by Chavez (1996). Tasseled cap values were calculated using coefficients described by Crist and Cicone (1984). Landsat bands are displayed with associated wavelengths (in micrometers).

possibility of directly monitoring understory removals in stands in which the overstory is unaltered.

HARVEST METRICS

Coppin et al. (2004) suggested that any approach to digital change detection includes both a change extraction algorithm and a set of rules for separating or labeling meaningful levels of change. These authors described ten categories of change extraction routines, the most common of which were postclassification comparison (*delta classification*), multidate composite analysis, univariate image differencing, image ratioing, and change vector analysis. Milne (1988), Singh (1989), and Coops et al. (Chapter 2, this volume) also discussed and categorized methods for detecting ecological change. The use of any of these processes depends on the identification of a measurement variable that is specifically relevant to the type of change studied.

Considerable variation exists in how harvests are labeled. The first section of this chapter described a hierarchy of harvest types based on silvicultural principles. However, vernacular usage of many silvicultural terms is sometimes inconsistent. Since the mapping process requires the identification of an attribute that is at least internally consistent, care should be taken in choosing a harvest measure that is both useful and unambiguous. This section focuses on how the effects of harvest may be conceptualized in a mapping context. It should be emphasized that all of the harvest variables to be discussed can be mapped with a range of levels of precision, depending on the quality of the supporting data and the ability of the chosen change extraction algorithm to identify necessary distinctions. Levels of precision may include simple detection of the presence/absence of harvest activity, classification of harvests into two or more categories, or estimating harvest effects as a continuous variable.

There are at least three ways that the changes resulting from harvest may be mapped: in terms of the general intensity of canopy removal; as a change in a particular biophysical variable (e.g., biomass, cover, density); or by silvicultural harvest type. The majority of published harvest detection studies identified harvests that remove most of the canopy; these harvests are alternately called forest cutovers (Hall et al., 1989), clearcuts (Banner and Ahern, 1995), and stand-replacing harvests (Cohen et al., 2002). Such studies typically create a binary map identifying cuts from specific time periods. Many such efforts lack either the reference data or the image spatial resolution to ensure a consistent silvicultural definition of the cuts that are mapped. However, any retention harvests that may be loosely classified as clearings are likely to resemble clearcuts in terms of percentage canopy removal. Thus, in practice, the attribute of interest in these maps is general canopy reduction. As a broad measure of harvest level, this attribute has value both in the monitoring of industrial management activities (Sader and Winne, 1992) and as an element of large-scale ecological processes (Cohen et al., 1996).

More specific quantification of the canopy removal resulting from harvest requires a more precise framework for describing the canopy. Potentially appropriate variables may include leaf area index, canopy cover, volume, basal area, and biomass. A variety of studies have shown these variables to be appropriate for characterization with remotely sensed data under certain conditions (Franklin, 2001; Hall et al., 1998; Lu et al., 2004; Mallinis et al., 2004; Steininger, 2000). Healey et al. (2006) estimated the changes in basal area and canopy cover over a range of harvest intensities using Landsat imagery, sample plots, and historical aerial photos. In addition to increasing the precision with which harvests can be labeled, modeling changes in specific biophysical variables expresses harvest in terms that may be of direct use in updating stand inventory information.

Another way to describe harvest is by silvicultural practice. As mentioned, limitations in the horizontal and vertical spatial resolution of many types of remotely sensed data realistically preclude direct monitoring of many of the silvicultural distinctions between harvest types. However, Franklin et al. (2000) and Sader et al. (2003) used harvest records to enable distinction between clearcuts and partial cuts.

In both studies, "clearcut" classes met the silvicultural definition of a clearcut, while all other harvests, including practices such as geometric harvesting and selective thinning, were grouped into the "partial cut" class. The advantage of considering harvests in silvicultural terms, particularly if individual partial cut practices can be identified, is that different practices may reflect different managerial priorities and create a richer context from which to project future developments.

As partial harvests become more common because of efforts to reduce fuel loads (Brown et al., 2004) and as a result of changing forest practice laws (Sader et al., 2003), the need for more nuanced harvest information will increase. Different metrics of forest removal have the ability to emphasize different structural or silvicultural elements of harvest. Therefore, the decision of how to label and differentiate harvest types should be based on monitoring needs and should be integrated into the design of the overall change detection approach.

CASE STUDY IN HARVEST DETECTION: STAND-REPLACING HARVESTS IN THE PACIFIC NORTHWEST (UNITED STATES)

BACKGROUND AND METHODS

The Northwest Forest Plan (the plan) was a 1994 amendment to the management plans of federal forestlands within the range of the spotted owl (*Strix occidentalis caurina*) in California, Oregon, and Washington. The aim of this plan was to balance the maintenance and restoration of older forest ecosystems with a predictable and sustainable level of harvest. Effectiveness monitoring was an important component of the plan, and separate monitoring programs were established to assess, among other plan outcomes, trends in old-growth forest ecosystems, habitat of spotted owls and marbled murrelets (*Brachyramphus marmoratus*), and the socioeconomic status of people in timber-dependent towns. Although a system of remeasured inventory plots has provided regionwide estimates of the net loss or gain of different forest types in the region, a spatially explicit record of significant disturbances was needed to assess changes to older forests and to owl habitat (Lint, 2005; Moeur et al., 2005). Furthermore, historical context was desired regarding harvest rates both before and after the plan was enacted.

The monitoring strategy designed to meet these needs included the mapping of stand-replacing harvests and fires in Oregon and Washington from 1972 to 2002 in approximately four-year intervals using composite analysis with Landsat MSS, TM, and Enhanced Thematic Mapper Plus (ETM +) data. Change detection in California was approached differently (Levien et al., 2002) and is not discussed here. Methods used in Oregon and Washington were chosen in consideration of project needs. Landsat imagery was used as it satisfied the need for historical data and because its moderate spatial resolution struck a balance between a large study area (14.5 million Ha; see Figure 3.4) and the need for accurate spatial referencing of disturbances. Also, the spectral resolution of the Landsat satellites, particularly TM and ETM+, which provide SWIR information, has been useful in several vegetation mapping projects in the region (Cohen and Goward, 2004). The mapping interval minimized

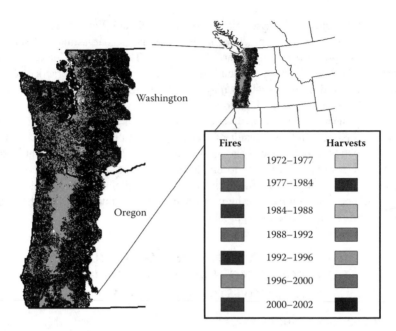

FIGURE 3.4 (See color insert following page 146.) Map of stand-replacing harvests and fires within the range of the northern spotted owl in Oregon and Washington from 1972 to 2002.

image acquisition and processing costs while offering sufficient temporal resolution for the detection of clearings in the Pacific Northwest (Cohen et al., 1998).

The stand-replacing harvest attribute (the detection of fires is not discussed here) met the study's need to identify cuttings that removed all or nearly all of a stand's trees. It should be noted that, silvicultural definitions notwithstanding, the stand-replacing designation used here does not apply to gradual shelterwood cuts that leave a large percentage of resident trees. In fact, harvests leaving a large canopy component (partial harvests) were intentionally excluded from the map. The identification and labeling of harvested pixels was accomplished through the use of composite analysis, a process by which a multitemporal "stack" of image data is classified to identify relevant changes. Accurate cross-date spatial coregistration is essential in this process (Coppin and Bauer, 1996), and an automated tie-point program (Kennedy and Cohen, 2003) was used to carefully coregister imagery. Other data preparation, detailed by Healey et al. (in review) included masking out nonforest areas using a land cover layer prepared for the plan area (Weyermann and Fassnacht, 2000) and subsetting Landsat images along general ecosystem boundaries to reduce ecological variation in the spectral signal. Composite analysis was chosen because it was judged to be an accurate and efficient means of isolating pixels displaying multitemporal spectral signatures consistent with stand-replacing harvest (Cohen and Fiorella, 1998). Supervised classification, by which spectral properties of disturbed and undisturbed pixels are identified in advance, was chosen for this analysis. An informal study (Figure 3.5) indicated that such a process, when used with a maximum likelihood classifier, is more efficient at isolating clearcut pixels than

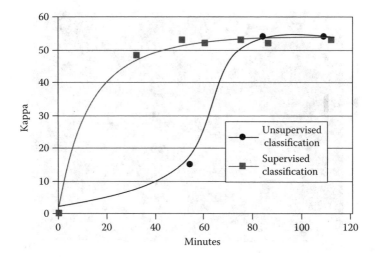

FIGURE 3.5 Pilot study of efficiency of composite analysis using unsupervised and supervised classification to detect stand-replacing harvests. Both approaches were applied in a 500,000-Ha study area in central Washington to create maps identifying stand-replacing harvests over two 4-year intervals using tasseled cap-transformed Landsat TM data. Resulting maps were evaluated against the Landsat imagery, allowing iterative adjustment of classification parameters (indicated by individual points). Kappa accuracy of successive maps was measured against hand-digitized disturbance maps and plotted over processing time. Reported times do not reflect preprocessing procedures that were common to both approaches.

unsupervised classification. Unsupervised classification relies on analyst interpretation of feature space clusters and can be time consuming when clusters are not well aligned with the boundaries of desired classes.

Prior to composite analysis, the tasseled cap transformation was performed on the Landsat data for data reduction and feature emphasis (see Data Considerations section). A further transformation, called the disturbance index (DI), was performed on the tasseled cap indices to produce a single band per image/date. This index quantifies the degree to which a pixel fits a profile that is presumed to match the position of clearcuts in tasseled cap space. Specifically, pixels with high tasseled cap brightness and low tasseled cap greenness and wetness values have high DI values. Details of the transformation can be found in the work of Healey et al. (2005). Pixels that move suddenly from low to high DI values are identified by the classifier as having been disturbed. Composite analysis in different regions has shown DI-transformed imagery to produce results comparable to tasseled cap-transformed data (Healey et al., 2005). Further, reduction to a single band allows visualization of multitemporal imagery in a single computer monitor. Figure 3.6 shows a three-date display of DI imagery that can be interpreted using additive color logic to identify the timing of each harvest. This ready identification of the date of each clearing facilitated the selection of training polygons for supervised composite analysis. Postprocessing of the Oregon and Washington disturbance map included the mosaicking of all of the mapped segments together and the removal of all mapped cuts

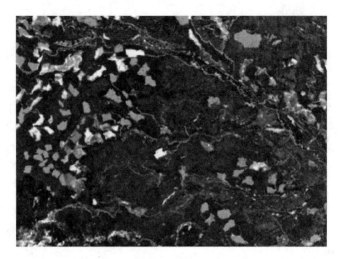

FIGURE 3.6 (See color insert following page 146.) Three dates of DI as viewed in a typical red-green-blue (RGB) monitor. The first date (1988) is plotted in the red color gun, the second (1992) in the green, and the third (1996) in the blue. Using the assumption that DI is high in disturbed areas, additive color logic can be used to interpret this multitemporal image. Cyan-colored areas are high in both the second and third dates, suggesting a disturbance between the first and second dates. Blue pixels have a high DI only in the third date, indicating the occurrence of a disturbance between the second and third dates. Reddish colors indicate stands disturbed prior to the first date that are becoming revegetated by the second and third dates.

and "islands" of retained trees that were less than 2 Ha in size. The latter measure was intended to remove small areas of error introduced by spatial misregistration.

RESULTS AND ANALYSES

The map that was created through this process (Figure 3.4) displayed stand-replacing harvests larger than 2 Ha and identified the time period in which they occurred. Map error was assessed at approximately 2500 randomly selected points through a sampling strategy described by Cohen et al. (2002). Error rates (88.9% overall accuracy with a kappa coefficient of 0.83), reported by Healey et al. (2006), were acceptable for the analyses described in the next paragraph. In general, the earlier dates, which were mapped with lower spatial resolution Landsat MSS data, were less accurate than TM-mapped dates. On the pixel level, most errors resulted from either spatial misregistration or confusion of partial harvests with stand-replacing harvests.

Three types of analyses were performed with the map in support of the plan's vegetation monitoring program. First, the map was considered in conjunction with a variety of geographical information system layers to identify harvest trends over ecological provinces, ownership categories, and federal land use designations. Figure 3.7 shows harvest trends by ownership. Notable in the graph is the dramatic reduction in harvest on federal Forest Service and Bureau of Land Management lands during the time that has coincided with the Northwest Forest Plan (Moeur et al., 2005). The second type of analysis focused on the average size of harvest units for each

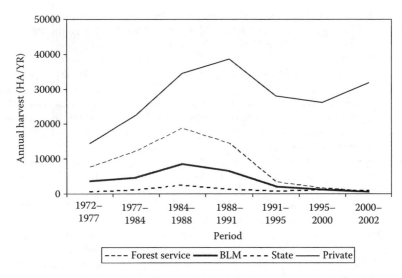

FIGURE 3.7 Annual area of stand-replacing harvest by major forestland owners in the Oregon portion of the Northwest Forest Plan area.

ownership over time; in general, stand-replacing harvests on private land were larger than those on public land over all time periods (Healey et al., in review). Finally, the map was combined with circa 1996 maps of older forests and owl and murrelet habitat to identify recent changes in those resources. Summaries of these analyses can be found in the work of Moeur et al. (2005) and Lint (2005).

CONCLUSIONS

Forest harvests can be measured in several ways. A large number of projects have focused on mapping clearcut-like operations that remove a majority of trees. Harvests can also be mapped according to silvicultural labels or as changes in specific biophysical variables. Each of these metrics can be mapped with a range of specificity, demanding more or less spatial, temporal, and spectral resolution of the supporting remotely sensed data set. Independent of the choice of detection algorithm, familiarity with expected harvest practices can facilitate interpretation of remotely sensed data for harvest mapping.

REFERENCES

Achard, F., Eva, H.D., Stibig, H.-J., Mayaux, P., Gallego, J., Richards, T., & Malingreau, J-P. (2002). Determination of deforestation rates of the world's humid tropical forests. *Science, 297*, 999–1002.

Alves, D.S., Pereira, J.L.G., de Souza, C.L., Soares, J.V., and Yamaguchi, F. (1999). Characterizing landscape changes in central Rondonia using Landsat TM imagery. *International Journal of Remote Sensing, 20*, 2877–2882.

Antikidis, E., Arino, O., Laur, H., and Arnaud, A. (1998). Deforestation evaluation by synergetic use of ERS SAR coherence and ATSR hot spots: the Indonesian fire event of 1997. *Earth Observation Quarterly, 59,* 34–38.

Banner, A.V. and Ahern, F.J. (1995). Forest clearcut mapping using airborne C-band SAR and simulated RADARSAT imagery. *Canadian Journal of Remote Sensing, 21,* 124–137.

Banner, A.V. and Lynham, T. (1981). Multitemporal analysis of Landsat data for forest cut over mapping — a trial of two procedures. *Proceedings of the Seventh Canadian Symposium on Remote Sensing* (pp. 233–240). Canadian Remote Sensing Society, Winnipeg, MB, Canada.

British Columbia, Ministry of Forests, Forest Practices Branch. (2003). *Silvicultural Systems Handbook for British Columbia.* Forest Practices Branch, BC Ministry of Forests, Victoria, BC, Canada. 208p.

Brown, R.T., Agee, J.K., and Franklin, J.F. (2004). Forest restoration and fire: principles in the context of place. *Conservation-Biology, 18,* 903–912.

Chavez, P.S. (1996). Image-based atmospheric corrections — revisited and improved. *Photogrammetric Engineering and Remote Sensing, 62,* 1025–1036.

Cihlar, J., Pultz, T.J., and Gray, A.L. (1992). Change detection with synthetic aperture radar. *International Journal of Remote Sensing, 55,* 153–162.

Clark, D.B., Castro, C.S., Alvarado, L.D.A., and Read, J.M. (2004). Quantifying mortality of tropical rain forest trees using high-spatial-resolution satellite data. *Ecology Letters, 7,* 52–59.

Cohen, W.B. and Fiorella, M. (1998). Comparison of methods for detecting conifer forest change with Thematic Mapper imagery. In R.S. Lunetta and C.D. Elvidge (Eds.), *Remote Sensing Change Detection: Environmental Monitoring Methods and Applications* (pp. 89–102). Ann Arbor Press, Chelsea, MI. 318p.

Cohen, W.B., Fiorella, M., Gray, J., Helmer, E., and Anderson, K. (1998). An efficient and accurate method for mapping forest clearcuts in the Pacific Northwest using Landsat Imagery. *Photogrammetric Engineering and Remote Sensing, 64,* 293–300.

Cohen, W.B. and Goward, S.N. (2004). Landsat's role in ecological applications of remote sensing. *Bioscience, 54,* 535–545.

Cohen, W.B., Harmon, M.E., Wallin, D.O., and Fiorella, M. (1996). Two decades of carbon flux from forests of the Pacific Northwest: estimates from a new modeling strategy. *Bioscience, 46,* 836–844.

Cohen, W.B., Spies, T.A., Alig, R.J., Oetter, D.R., Maiersperger, T.K., and Fiorella, M. (2002). Characterizing 23 years (1972–1995) of stand-replacing disturbance in western Oregon forest with Landsat imagery. *Ecosystems, 5,* 122–137.

Collins, J.B. and Woodcock, C.E. (1996). An assessment of several linear change detection techniques for mapping forest mortality using multitemporal Landsat TM data. *Remote Sensing of Environment, 56,* 66–77.

Coppin, P.R. and Bauer, M.E. (1996). Digital change detection in forest ecosystems with remote sensing imagery. *Remote Sensing Reviews, 13,* 207–234.

Coppin, P., Jonckheere, I., Nackaerts, K., Muys, B., and Lambin, E. (2004). Digital change detection methods in ecosystem monitoring: a review. *International Journal of Remote Sensing, 25,* 1565–1596.

Crist, E.P. and Cicone, R.C. (1984). A physically-based transformation of Thematic Mapper data — the TM Tasseled Cap. *IEEE Transactions on Geoscience and Remote Sensing, 22,* 256–263.

Curtis, J.M.R. and Taylor, E.B. (2004). The genetic structure of coastal giant salamanders (*Dicamptodon tenebrosus*) in a managed forest. *Biological Conservation, 115,* 45–54.

Danson, P.M., and Curran, P.J., (1993). Factors affecting the remotely sensed response of coniferous forest plantations. *Remote Sensing of Environment, 43*, 55–65.

Donovan, G.H. (2004). Consumer willingness to pay a price premium for standing-dead Alaska yellow cedar. *Forest Products Journal, 54*, 38–42.

Doruska, P.F. and Nolen, W.R., Jr. (1999). Use of stand density index to schedule thinnings in loblolly pine plantations: a spreadsheet approach. *Southern Journal of Applied Forestry, 23*, 21–29.

Drieman, J.A. (1994). Forest cover typing and clearcut mapping in Newfoundland with C-band SAR. *Canadian Journal of Remote Sensing, 20*, 11–16.

Ehrlich, D., Lambin, E.F., and Malingreau, J.P. (1997). Biomass burning and broad-scale land-cover changes in Western Africa. *Remote Sensing of Environment, 61*, 201–209.

Fiedler, C.E. (2004). *A Strategic Assessment of Crown Fire Hazard in Montana: Potential Effectiveness and Costs of Hazard Reduction Treatments.* U.S. Department of Agriculture, Forest Service, Pacific Northwest Research Station, Portland, OR. General Technical Report PNW 622. 48p.

Fight, R.D. (2004). *Thinning and Prescribed Fire and Projected Trends in Wood Product Potential, Financial Return, and Fire Hazard in New Mexico.* U.S. Department of Agriculture, Forest Service, Pacific Northwest Research Station, Portland, OR. General Technical Report PNW 605. 48p.

Fischer, C.S. and Levien, L.M. (2001). *Monitoring California's Hardwood Rangelands Using Remotely Sensed Data.* In: Standiford, R.B., McCreary, D., Purcell, K.L. (Eds.) *Proceedings of the Fifth Symposium on Oak Woodlands: Oaks in California's Changing Landscape, October 22–25.* pp 603–615. Gen. Tech. Report PSW-GTR-184. Albany, CA: PSW Research Station, Forest Service, Dept. Agriculture. 846p.

Franklin, S.E. (2001). *Remote Sensing for Sustainable Forest Management.* CRC Press/Lewis Publishers, Boca Raton. 407p.

Franklin, S.E., Moskal, L.M., Lavigne, M.B., and Pugh, K. (2000). Interpretation and classification of partially harvested forest stands in the Fundy Model Forest using multitemporal Landsat TM. *Canadian Journal of Remote Sensing, 26*, 318–333.

Gong, P. and Xu, B. (2003). Remote sensing of forests over time: change types, methods, and opportunities. In M. Wulder and S. Franklin (Eds.), *Methods and Applications for Remote Sensing: Concepts and Case Studies* (pp. 301–333). Kluwer Academic Publishers, Norwell, MA. 519p.

Hall, R.J., Franklin, S.E., and Gerylo, G.R. (1998). Estimation of stand volume from high resolution multispectral imagery. *Proceedings 20th Canadian Symposium on Remote Sensing* (pp. 191–196). Canadian Aeronautics and Space Institute, Ottawa, ON, Canada.

Hall, R.J., Kruger, A.R., Scheffer, J., Titus, S.J., and Moore, W.C. (1989). A statistical evaluation of Landsat TM and MSS for mapping forest cutovers. *Forest Chronicle, 65*, 441–449.

Hassett, J.E. and Zak, D.R. (2005). Aspen harvest intensity decreases microbial biomass, extracellular enzyme activity, and soil nitrogen cycling. *Soil Science of America, 69*, 227–235.

Hayes, D.J. and Sader, S.A. (2001). Comparison of change-detection techniques for monitoring tropical forest clearing and vegetation regrowth in a time series. *Photogrammetric Engineering and Remote Sensing, 67*, 1067–1075.

Healey, S.P., Cohen, W.B., Moeur, M., Spies, T.A., Whitley, M.G., Lefsky, M.A., and Pflugemacher, D. (In review). Evidence from the Landsat Satellite record that different harvest practices have created fundamental differences between Federal and private forests in the Pacific Northwest of the United States. J. Forestry.

Healey, S.P., Cohen, W.B., Yang, Z., and Krankina, O.N. (2005). Comparison of tasseled cap-based Landsat data structures for use in forest disturbance detection. *Remote Sensing Environment, 97,* 301–310.

Healey, S.P., Cohen, W.B., Yang, Z., and Pierce, D.J. (2006). Application of two regression-based methods to estimate the effects of partial harvest on forest structure using Landsat data. *Remote Sensing of Environment, 101,* 115–126.

Healthy Forests Restoration Act. (2003). Public Law No: 108-148. 29. Retrieved from: http://resourcescommittee.house.gov/issues/ffh/hfra_publaw.pdf on Jan. 1, 2006.

Heikkonen, J. and Varjö, J. (2004). Forest change detection applying Landsat Thematic Mapper difference features: a comparison of different classifiers in boreal forest conditions. *Forest Science, 50,* 579–588.

Jet Propulsion Laboratory. (1986). *Shuttle Imaging Radar-C Science Plan.* Jet Propulsion Laboratory, Pasadena, CA. JPL Publication 86-29.

Jin, S.M. and Sader, S.A. (2005). Comparison of time series tasseled cap wetness and the normalized difference moisture index in detecting forest disturbances. *Remote Sensing of Environment, 94,* 364–372.

Kennedy, R.E. and Cohen, W.B. (2003). Automated designation of tie-points for image-to-image coregistration. *International Journal of Remote Sensing, 24,* 3467–3490.

Kline, J.D., Azuma, D.L., and Alig, R.J. (2004). Population growth, urban expansion, and private forestry in western Oregon. *Forest Science, 50,* 33–43.

Lambin, E.F. and Ehrlich, D. (1997). Land-cover changes in sub-Saharan Africa (1982–1991): application of a change index based on remotely sensed surface temperature and vegetation indices at a continental scale. *Remote Sensing of Environment, 61,* 181–200.

Lefsky, M.A. and Cohen, W.B. (2003). Selection of remotely sensed data. In M. Wulder and S. Franklin (Eds.), *Methods and Applications for Remote Sensing: Concepts and Case Studies* (pp. 13–46). Kluwer Academic Publishers, Norwell, MA. 519p.

Levien, L., Fischer, C., Roffers, P., Maurizi, B., and Suero, J. (2002). *Monitoring Land Cover Changes in California.* California Land Cover Mapping and Monitoring Program. Northeastern California Project Area. USDA Forest Service and California Department of Forestry and Fire Protection Cooperative Monitoring Program. 171p. State of California, Resources Agency, Department of Forestry and Fire Protection, Sacramento, CA.

Lint, J. (2005). *Northwest Forest Plan — The First 10 Years (1994–2003): Status and Trends of Northern Spotted Owl Populations and Habitat.* U.S. Department of Agriculture, Forest Service, Pacific Northwest Research Station, Portland, OR. Gen. Tech. Rep. PNW-GTR-648. 176p.

Lu, D., Mausel, P., Brondizio, E., and Moran, E. (2004). Relationships between forest stand parameters and Landsat TM spectral responses in the Brazilian Amazon basin. *Forest Ecology and Management, 198,* 149–167.

Luckman, A., Baker, J., Honzak, M., and Lucas, R. (1998). Tropical forest biomass density estimation using JERS-1 SAR: seasonal variation, confidence limits, and application to image mosaics. *Remote Sensing of Environment, 63,* 126–139.

Lunetta, R.S., Lyon, J.G., Guindon, B., and Elvidge, C.D. (1998). North American landscape characterization: "triplicate" data sets and data fusion products. In R.S. Lunetta and C.D. Elvidge (Eds.), *Remote Sensing Change Detection: Environmental Monitoring Methods and Applications* (pp. 41–52). Ann Arbor Press, Chelsea, MI. 318p.

Maguire, D. (2005). Uneven-aged management: panacea, viable alternative, or component of a grander strategy? *Journal of Forestry, 103,* 73–74.

Mallinis, G., Koutsias, N., Makras, A. and Karteris, M. (2004). Forest parameters estimation in a European Mediterranean landscape using remotely sensed data. *Forest Science, 50,* 450–460.

Masek, J. (2005). *LEDAPS Disturbance Index: Algorithm Description v.1.* Algorithm Description for LEDAPS disturbance products. Retrieved from http://ledaps.nascom.nasa. gov/ledaps/docs1.html on Jan. 1, 2006.

Miller, L.D., Nualchawee, K., and Tom, K. (1978). *Analysis of the Dynamics of Shifting Cultivation in the Tropic Forest of Northern Thailand Using Landscape Modeling and Classification of Landsat Imagery.* NASA Goddard Space Flight Center, Greenbelt, MD. Technical Memorandum No. 79545.

Milne, A.K. (1988). Change detection analysis using Landsat imagery: a review of methodology. *Proceedings of IGARSS'88 Symposium* (pp. 541–544). Edinburgh, Scotland. ESA, Noordwijk, The Netherlands. ESA SP-284.

Moeur, M., Spies, T.A., Hemstrom, M., Martin, J.R., Alegria, J., Browning, J., Cissel, J., Cohen, W.B., Demeo, T.E., Healey, S., and Warbington, R. (2005). *Northwest Forest Plan — The First 10 Years (1994–2003): Status and Trend of Late-Successional and Old-Growth Forest.* U.S. Department of Agriculture, Forest Service, Pacific Northwest Research Station, Portland, OR. Gen. Tech. Rep. PNW-GTR-646. 142p.

Nilson, T., Olsson, H., Anniste, J., Lükk, T., and Praks, J. (2001). Thinning-caused change in reflectance of ground vegetation in boreal forest. *International Journal of Remote Sensing, 22,* 2763–2776.

Oliver, C.D. (1981). Forest development in North America following major disturbance. *Forest Ecology and Management, 3,* 153–168.

Oliver, C.D. and Larson, B.G. (1996). *Forest Stand Dynamics.* John Wiley and Sons, New York. 537p.

Olsson, H. (1994). Changes in satellite-measured reflectances caused by thinning cuttings in boreal forest. *Remote Sensing of Environment, 50,* 221–230.

Potter, C., Tan, P.N., Kumar, V., Kucharik, C., Klooster, S., Genovese, V., Cohen, W., and Healey S. (2005). Recent history of large-scale ecosystem disturbances in North America derived from the AVHRR satellite record. *Ecosystems, 8,* 808–824.

Ranson, K., Kovacs, K., Sun, G., and Kharuk, V. (2003). Disturbance recognition in the boreal forest using radar and Landsat-7. *Canadian Journal of Remote Sensing, 29,* 271–285.

Read, J.M., Clark, D.B., Venticinque, E.M., and Moreira, M.P. (2003). Application of merged 1-m and 4-m resolution satellite data to research and management in tropical forests. *Journal of Applied Ecology, 40,* 592–600.

Richards, W.H., Wallin, D.O., and Schumaker, N.H. (2002). An analysis of late-seral forest connectivity in Western Oregon, U.S.A. *Conservation Biology, 16,* 1409–1421.

Rigina, O. (2003). Detection of boreal forest decline with high-resolution panchromatic satellite imagery. *International Journal of Remote Sensing, 24,* 1895–1912.

Saatchi, S., Soares, J.B., and Alves, M. (1997). Mapping deforestation and land use in Amazon rainforest by using SIR-C imagery. *Remote Sensing of Environment, 59,* 191–202.

Sader, S.A., Bertrand, M., and Wilson, E.H. (2003). Satellite change detection of forest harvest patterns on an industrial forest landscape. *Forest Science, 49,* 341–353.

Sader, S.A. and Winne, J.C. (1992). RGB-NDVI colour composites for visualizing forest change dynamics. *International Journal of Remote Sensing, 13,* 3055–3067.

Saksa, T., Uuttera, J., Kolström, T., Lehikionen, M., Pekkarinen, A., and Sarvi, V. (2003). Clear-cut detection in boreal forest aided by remote sensing. *Scandinavian Journal of Forest Research, 18,* 537–546.

Sanden, J.J. van-der, and Hoekman, D.H. (1999). Potential of airborne radar to support the assessment of land cover in a tropical rain forest environment. *Remote Sensing of Environment, 68,* 26–40.

Singh, A. (1989). Digital change detection techniques using remotely sensed data. *International Journal of Remote Sensing, 10,* 989–1001

Skakun, R.S., Wulder, M.A., and Franklin, S.E. (2003). Sensitivity of Thematic Mapper enhanced wetness difference index to detect mountain pine beetle red-attack damage. *Remote Sensing of Environment, 86*, 433–443.

Skole, D., and Tucker, C. (1993). Tropical deforestation and habitat fragmentation in the Amazon: satellite data from 1978 to 1988. *Science, 260*, 1905–1910.

Smith, D.M., Larson, B.C., Kelty, M.J., and Ashton, P.M.S. (1997). *The Practice of Silviculture: Applied Forest Ecology.* 6th ed. John Wiley and Sons, New York. 537p.

Smith, G. and Askne, J. (1997). Use of ERS SAR interferometry for monitoring clear-cutting of forests. *Proceedings of IGARSS'97 Symposium, Remote Sensing — A Scientific Vision for Sustainable Development* (pp. 793–796). Singapore, August 3–8, 1997. IEEE, Piscataway, NJ.

Smith, G. and Askne, J. (2001). Clear-cut detection using ERS interferometry. *International Journal of Remote Sensing, 22*, 3651–3664.

Steininger, M.K. (2000). Satellite estimation of tropical secondary forest above-ground biomass: data from Brazil and Bolivia. *International Journal of Remote Sensing, 21*, 1139–1157.

Swank, W.T., Vose, J.M., Elliott, K.J., Brooks, R.T., and Lust, N. (2001). Long-term hydrologic and water quality responses following commercial clearcutting of mixed hardwoods on a southern Appalachian catchment. *Forest Ecology and Management, 143*, 163–178.

Treuhaft, R.N., Law, B.E., and Asner, G.P. (2004). Forest attributes from radar interferometric structure and its fusion with optical remote sensing. *BioScience, 54*, 561–571.

Varjö, J. (1996). Controlling continuously updated forest data by satellite remote sensing. *International Journal of Remote Sensing, 17,* 43–67.

Wegmüller, U. and Werner, C.L. (1995). SAR interferometric signatures of forest. *IEEE Transactions on Geoscience and Remote Sensing, 33,* 1153–1161.

Weyermann, D. and Fassnacht, K. (2000). The interagency vegetation mapping project: estimating certain forest characteristics using Landsat TM data and forest inventory plot data. In J.D. Greer (Ed.), *Remote Sensing and Geospatial Technologies for the New Millennium.* Proceedings of the Eight Forest Service Remote Sensing Applications Conference (CD-ROM). Mira CD-ROM Publishing, St. Louis, MO. 11p.

Williams, D.L. and Nelson, R.F. (1986). Use of remotely sensed data for assessing forest stand conditions in the eastern United States. *IEEE Transactions on Geoscience and Remote Sensing, 24*, 130–138.

Wilson, J.S. and Oliver, C.D. (2000). Stability and density management in Douglas-fir plantations. *Canadian Journal of Forest Research, 30*, 910–920.

Woodcock, C.E., Macomber, S.A., Pax-Lenney, M., and Cohen, W.B. (2001). Monitoring large areas for forest change using Landsat: generalization across space, time and Landsat sensors. *Remote Sensing of Environment, 78*, 194–203.

Woodcock, C.E. and Strahler, A.H. (1987). The factor of scale in remote sensing. *Remote Sensing of Environment, 21*, 311–332.

Yang, Z. (2004). Early forest succession following clearcuts in western Oregon: patterns and abiotic controls. Ph.D. dissertation. Oregon State University, Department of Forest Science, Corvallis, OR. 101p.

Yatabe, S.M. and Leckie, D.C. (1995). Clearcut and forest-type discrimination in satellite SAR imagery. *Canadian Journal of Remote Sensing, 21*, 455–467.

Zhan, X., Sohlberg, R., Townshend, J.R.G., DiMiceli, C., Carroll, M., Eastman, J.C., Hansen, M.R.S., and DeFries, R.S. (2002). Detection of land cover changes using MODIS 250m data. *Remote Sensing of Environment, 83*, 336–350.

4 Remotely Sensed Data in the Mapping of Insect Defoliation

*Ronald J. Hall, Robert S. Skakun,
and Eric J. Arsenault*

CONTENTS

INTRODUCTION

There is an increasing need by government agencies and industry to map and monitor the area, severity, and spatial location of insect defoliation consistently to report and assess its impact on forest health and productivity. Information about pest activity is used to prescribe appropriate pest management practices and to measure the sustainability of forest ecosystems from timber supply and nontimber value perspectives (P. J. Hall and Moody, 1994; R. J. Hall et al., 2003; MacLean, 1990; Simpson and Coy, 1999). Damage from insect defoliation will (a) have an impact on trees and stands by causing timber volume changes due to mortality and growth loss; (b) influence host–pest interactions, including predisposition to secondary host infection; and (c) cause direct changes to stand dynamics (Alfaro, 1988; Coulson and Witter, 1984; Ives and Wong, 1988; Kulman, 1976). Aerial sketch mapping has been the most frequently used technique to map insect defoliation of North American forests (Ciesla, 2000; Harris and Dawson, 1979; Simpson and Coy, 1999). While the value of long-standing records is without question, there are limitations regarding the extent that aerial surveys can be used to relate defoliated area (the mapped quantity) to impact (the quantity of growth and wood volume loss) (MacLean, 1990). The questions of when, where, and how much damage remain fundamental to forest health concerns. Of interest is determining the role that remote sensing may play in providing answers to some of these questions.

Insect defoliation affects the morphological and physiological characteristics of trees, and it is these characteristics that govern how trees absorb and reflect light (Murtha, 1982). The remote sensing approach has been to relate differences in spectral response to chlorosis (yellowing), foliage reddening, or foliage reduction over time, assuming that these differences can be interpreted, classified, or correlated to damage caused by insect activity (Franklin, 2001; R. J. Hall et al., 1983). Remote sensing data has long been explored for detecting and mapping insect defoliation; variable success has been reported (Dottavio and Williams, 1983; Franklin, 2001; Franklin and Raske, 1994; R. J. Hall et al., 2003; Heikkilä et al., 2002; Leckie and Ostaff, 1988; Radeloff et al., 1999; Riley, 1989; Royle and Lathrop, 1997). The range of remote sensing applications has included detecting and mapping defoliation, characterizing patterns of disturbance, modeling and predicting outbreak patterns, and providing data to pest management decision support systems. The possibility of forecasting the susceptibility and vulnerability of forested areas to insect defoliation has also been reported as a tool to provide mitigation options to forest managers (Luther et al., 1997). These applications were intended to produce information products that support pest management planning, impact studies, and regional or national reporting. A difficult challenge has been the use of visual or optical estimates of defoliation severity to define a spectral basis for damage class limits that can be mapped from the remote sensing image. This is an important problem requiring resolution if consistent detection and mapping is to be achieved (R. J. Hall et al., 2003). As a result, despite past research and apparent high potential for use of remote sensing to map insect defoliation, it remains a technology that has seen relatively little operational use (D. J. Peterson et al., 1999).

Successful use of remote sensing for entomological studies requires integrating knowledge about the insect pest, forest host, and remotely sensed image. Reliance on the remotely sensed image alone is insufficient because what is observed is the manifestation of damage rather than the causal agent itself. Knowledge of insect pest biology and its manifestation of damage can be related to species host phenology, stand composition, and structure to understand its damage impact. In turn, this knowledge is fundamental for remote sensing with respect to defining spectral regions appropriate for damage assessment, determining sensor spatial resolution requirements, identifying the optimum timing for data acquisition, and selecting or developing image-processing methods for mapping pest damage. What are the major defoliators in North America from which this integrative framework could be employed?

Based on a review of forest health reports in Canada and the continental United States, six insect pests are considered among the major defoliators of deciduous and coniferous forests of North America. These insect pests include aspen defoliators such as the forest tent caterpillar (*Malacosoma disstria* Hubner) and large aspen tortrix (*Choristoneura conflictana* Wlk.); gypsy moth (*Lymantria dispar* L.); spruce budworm (*Choristoneura fumiferana* [Clem.] in the east and *Choristoneura occidentalis* Freeman in the west); eastern hemlock looper (*Lambdina fiscellaria fiscellaria* [Guen.]); and jack pine budworm (*Choristoneura pinus pinus* Freeman) (Hall et al., 1998; Simpson and Coy, 1999; U. S. Department of Agriculture [USDA], 2004). Many of these pests can cause periodic outbreaks over large areas that can culminate in extensive replacement of forest stands (Volney and Fleming, 2000). This chapter has been written within the context of these major North American defoliators.

The purposes of this chapter are to:

1. Compare the general biological characteristics of six insect defoliators, their damage patterns, and the timing of defoliation damage
2. Summarize defoliation mapping methods, including aerial sketch map surveys, and which remotely sensed data and methods have been applied to these six insect defoliators
3. Present a case study for mapping and monitoring aspen defoliation over a multitemporal sequence of four image dates

MAJOR INSECT DEFOLIATORS IN NORTH AMERICA

The following sections include, for select major insect defoliators in North America, brief biological descriptions in relation to defoliation damage, timing, and remotely sensed image acquisition considerations.

ASPEN DEFOLIATORS: FOREST TENT CATERPILLAR AND LARGE ASPEN TORTRIX

The forest tent caterpillar and large aspen tortrix are the most serious insect defoliators of trembling aspen in North America (Figure 4.1a). These defoliators have affected large areas of forests in Canada and the continental United States, with a

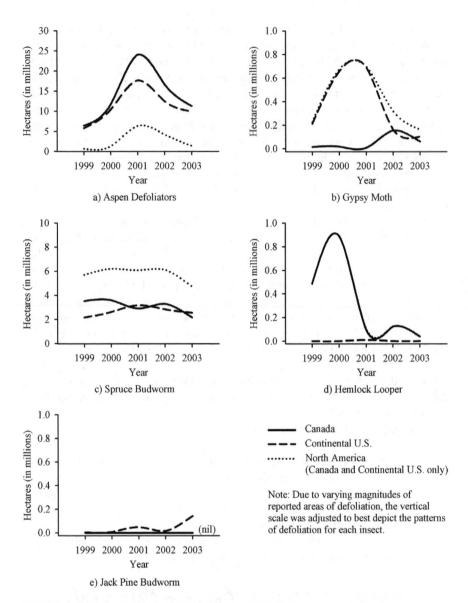

FIGURE 4.1 (See color insert following page 146.) Patterns of moderate-to-severe insect defoliation from 1999 to 2003 in Canada and the continental United States.

notable peak in 2001 that approached 25 million Ha (Figure 4.1a; USDA, 2004). Damage is caused during larval feeding, which begins when the leaves emerge in the spring and lasts until the end of June or early July depending on the defoliator (Table 4.1). Trembling aspen tends to refoliate during late July depending on the duration and severity of defoliation. The ideal time for observing the actual severity of defoliation is when the largest amount of foliage has been consumed. This would be near the culmination of larval feeding, which occurs approximately from mid-

June to early July. This confines the time period when remote sensing imagery should be acquired. If the severity of defoliation is known to be severe based on field surveys or prior aerial survey knowledge, then it may be possible to obtain after-defoliation imagery earlier in June during the larval feeding period under the assumption that both before- and after-defoliation imagery would be acquired since the intensity of defoliation would be sufficiently persistent and dramatic to permit a wider time window to be used (R. J. Hall et al., 1983). While there was some reported success at interpreting and classifying the presence or absence of severe aspen defoliation from multidate Landsat images (R. J. Hall et al., 1983, 1984), more recent work has resulted in greater sensitivity to defoliation severity levels through detection of differences in leaf area index (LAI) between pre- and postoutbreak images (R. J. Hall et al., 2003).

GYPSY MOTH

In 1869, gypsy moth was introduced into the Boston area from France for experimental crossbreeding with silkworms (Leatherman et al., 1995). Some escaped, and without natural enemies, they have become the most important defoliator of trees in the eastern United States (Herms and Shetlar, 2002). The gypsy moth was observed in eastern Canada in 1981 as a primary defoliator of red oak, although it has also defoliated white birch, red maple, and eastern white pine (Hall et al., 1998). During 1999 to 2001, it was a more significant defoliator in the United States compared to Canada, but its area of damage was more similar in 2003 (Figure 4.1b). Gypsy moth larvae emerge in late April to early May and begin feeding immediately, although the major defoliation damage occurs from older, larger larvae during early to mid-June (Table 4.1). Because this pest has such a wide host base that includes deciduous and coniferous species, its detection and mapping by remote sensing can be challenging. As a result, a multitude of remote sensing techniques, including band ratios, supervised and unsupervised classifications, image differencing, and change vector analysis, to name a few, has been employed in gypsy moth studies (Hurley et al., 2004; Joria et al., 1991; Muchoney and Haack, 1994; Townsend et al., 2004). While refoliation in deciduous stands is often considered a bounding condition on image selection windows, Hurley et al. (2004) used three dates of image ratios representing before-defoliation, after-defoliation, and refoliation time periods to increase circumstantial evidence in the detection of gypsy moth defoliation events.

SPRUCE BUDWORM

The eastern and western spruce budworms differ little in biology and are among the most destructive defoliators in North America (MacLean, 1990; Volney, 1985). The total area of defoliation in Canada and the United States has been relatively similar, within the range of 2 to 4 million Ha per year (Figure 4.1c). Duration of outbreaks appears to vary, but its damage pattern frequently includes current-year and older foliage and results in growth loss, top kill, and vast areas of mortality, especially if severe defoliation is repeated over several years (MacLean, 1990; Ostaff and MacLean, 1989). Larval feeding begins with emergence of the overwintering second

TABLE 4.1
Comparative Biology of Six Major Deciduous and Coniferous Defoliators

Species	Egg stage	Larval stage	Pupal stage	Adult stage	Feeding preference	Preferred hosts	Optimal habitat	References
Aspen defoliator: forest tent caterpillar	Overwinters in egg bands	5 instars, emerges in spring, feeds 5–6 weeks to early July	Mid- to late July; adults emerge in 10 days	Late July to early August	Aspen foliage	Trembling aspen, other deciduous species	No apparent preference although moisture stress and environmental conditions may increase damage	Ives and Wong (1988); E. B. Peterson and Peterson (1992); Cerezke and Volney (1995)
Aspen defoliator: large aspen tortrix	Laid from mid-June to early July; hatches in 2 weeks during early to late July	5 instars, overwinters as second instar under bark scales; emerges when buds start to swell	Early to mid-June; adults emerge in 10 days	7–14 days following pupal stage in mid-June to early July	Aspen foliage	Trembling aspen		Furniss and Carolin (1977); Ives and Wong (1988); E. B. Peterson and Peterson (1992)
Gypsy moth	Overwintering egg mass on many substrates, emergence in late April to early May	6 instars female, 5 instars male; emerge in spring; feeds May to June	Early to midsummer, typically July, approximately 2 weeks	Late spring or early summer	Largely a deciduous defoliator, may feed on conifers	Many hosts: alder, aspen, birch, oak, tamarack, willow	No apparent preference for specific habitat conditions although overmature, suppressed, diseased trees, drought increase hazard	Nealis and Erb (1993); Jobin (1995); Leatherman et al. (1995)

Insect							References
Spruce budworm	July–August	6 instars, *fumiferana*: overwintering larvae emerges as 2nd instar in late April – May *occidentalis*: emerges in spring, feeds to late June	*fumiferana*: June *occidentalis*: late June to early July	*fumiferana*: late June to July; mainly adult dispersal *occidentalis*: emerges 10 days following pupation	Male flowers, young foliage	Many hosts: *fumiferana*: balsam fir, white spruce, black spruce *occidentalis*: Douglas fir, grand fir, white fir, subalpine fir	Warm temperatures, dry sites, dense uneven-aged mature stands	Fellin and Dewey (1982); Martineau (1984); Ives and Wong (1988); Ostaff and MacLean (1989); Shepherd et al. (1995)
Hemlock looper	Overwintering egg laid from August to October	Emerges in May/June, 4 instars through to August	August to September in the soil/litter, bark crevices, moss on bark	September to October	Young foliage	Eastern hemlock, balsam fir, white spruce, many others	Stressed trees, climatic factors leading to decreased vigor of forest stands	Coulson and Witter (1984); Martineau (1984); Raske et al. 1995
Jack pine budworm	August – September, emerging 6–10 days after deposit	7 instars, spring to early July, second instar overwinters	Early to mid-July on branches	July to August; deposit eggs	Male staminate flowers, young foliage	Jack pine, Scots pine, red pine, lodgepole pine	Overstocked, overmature stands on poor-quality sites	Kulman et al. (1963); Ives and Wong (1988); Cadogan (1995)

instar larvae in the spring until feeding culmination in June (Table 4.1). Spruce budworm can defoliate several host tree species depending on its location and is compounded by current and cumulative defoliation conditions, which complicates detection by remote sensing. During the latter stages of larval feeding, the needles turn reddish brown, which is an indicator of current defoliation, but the detection window is narrow relative to other insect pests, which restricts the time period when remotely sensed imagery could be acquired. Research has instead focused on the reflectance characteristics of cumulative defoliation (Leckie et al., 1988). As a result, there has been greater relative success at detection of cumulative damage even without the use of multidate imagery (Franklin and Raske, 1994).

HEMLOCK LOOPER

The distribution of eastern hemlock looper extends from Alberta to Newfoundland in Canada and then south to Georgia in the eastern United States (MacLean and Ebert, 1999). While periodic large outbreaks have occurred in North America, the trend from 1999 to 2003 suggests it is a more significant pest in Canada (Figure 4.1d). The principal host is balsam fir, but other tree species become susceptible in high populations as the looper is an aggressive feeder that will consume foliage of all age classes (MacLean and Ebert, 1999; Raske et al., 1995). Emerging larvae in the spring feed on new needles, while later instars will feed on both new and old foliage; feeding is generally completed by late July (Table 4.1). Because the looper is a wasteful feeder, the residual foliage tends to get caught with the silken threads, and the desiccation of the damaged needles gives the trees their characteristic red-brown color. It is at this stage of damage that remote sensing studies tend to be focused (Franklin, 1989). Significant correlations have been reported between percentage defoliation and spectral reflectance values, particularly in the near- ($r = -0.78$, $p =.0001$) and middle-infrared ($r = -0.63$, $p =.0001$) portions of the spectrum, suggesting that discrimination among light, moderate, and severe classes of defoliation was possible from single-date Landsat Thematic Mapper (TM) data (Luther et al., 1991). Outbreaks of this pest can affect large areas, such as the 1998–2000 event that defoliated over 400,000 Ha and was detected and mapped from 1-km coarse spatial resolution multitemporal SPOT (Système pour l'Observation de la Terre) Vegetation data (Fraser and Latifovic, 2005).

JACK PINE BUDWORM

The jack pine budworm is considered the most important defoliator of jack pine in Canada and the lake states of the United States (Cadogan, 1995). Previous reports suggested outbreak patterns occur every 6 to 10 years (Volney and McCullough, 1994), when large areas can be damaged, and 1999–2003 trends suggest defoliation is currently at endemic levels (Figure 4.1e). Jack pine budworm overwinters as a second instar larva and emerges in late May soon after male cones open and new, young needles emerge (Table 4.1). The budworm larvae migrate to the tops and outer crown of trees due to their preference for male flower clusters and young foliage (Howse, 1984). Defoliation spreads from the top of the tree downward (Moody,

1986). Only the basal portion of the needle is eaten, while the rest becomes entangled in a mass of silk and larval excrement, which changes to a reddish color that becomes an indicator of defoliation severity (Volney, 1988). The red discoloration is likely the stage at which the greatest spectral change occurs, and as such, the timing for the mapping of defoliation is critical because peak coloration occurs during a short period from late June to early July (Howse, 1984). Wind and rain remove the red needles, resulting in exposed branches and top kill if severe defoliation was sustained. While the degree of top kill sustained in a stand is an indicator of defoliation severity (Hall et al., 1998), attempts to detect top kill on multidate satellite imagery proved difficult (R. J. Hall et al., 1995). The short time period during which the red discoloration is visible on trees remains the most appropriate time period for remote sensing observation (Leckie et al., 2005; Radeloff et al., 1999), but this does result in a very narrow window for acquiring cloud-free satellite images.

DISCUSSION: LINKING BIOLOGY WITH TECHNOLOGY

From the reviews of the biology and manifestation of damage caused by these forest pests (Table 4.1) and examples of remote sensing studies that have been undertaken, two issues appear fundamental to the successful use of remote sensing to map insect defoliation: the spectral and spatial characterization of defoliation and the timing of image acquisition. First, a remote sensing spectral basis for damage class limits (e.g., light, moderate, and severe) is required to achieve consistent detection and mapping of defoliation severity. Field and aerial surveys tend to rate areas defoliated into categories that remote sensing studies have attempted to emulate. Broad damage class limits are not conducive for consistent defoliation mapping because they may not correspond to differences in spectral response values that are spectrally or statistically separable on the image. The two factors that drive the spectral response of a sensor include its radiometric resolution and the range of sensitivity to the electromagnetic spectrum. Thus, remote sensing observations from airborne or satellite sensors are over a more continuous scale of spectral responses that can potentially capture a finer scale of defoliation levels than the broad classes that are typically used (Franklin, 2001; R. J. Hall et al., 2003).

Defoliation tends to result in either physical loss of leaf area or leaf color change, which results in physical differences in spectral response when compared to predefoliation images. Several consecutive years of defoliation, however, tend to result in physiological weakening, top kill, and mortality for some defoliators. Understanding the role these factors may play in the resulting spectral responses recorded in the image is important to successful use of remote sensing for mapping defoliation. In addition to the spectral observations of defoliation, the size of the outbreak area must also be large enough to be detectable with the airborne or satellite sensor employed. The spatial resolution of the sensor and the areal coverage of an image are also important considerations in the selection of the appropriate sensor. As a result, with both sensor spectral and spatial resolution considerations, the remote sensing of a defoliation problem is more complex than a simple change in foliage condition. Second, the timing of image data acquisition should coincide with the period when spectral changes resulting from defoliation are most observable; for many defoliators, this time period is often relatively short. Hardwoods, for example, tend to respond to defoliation by a second leaf flush in late spring or early summer (Radeloff et al., 1999). Red needles on conifers tend to be washed and blown away by rain and wind, leaving a predominantly green tree with only a reduction in needle leaf area that may or may not be detectable by remote sensing.

Timing of data acquisition is notably one of the most difficult to achieve with satellite remote sensing because of the need for cloud-free conditions during the suitable range of dates for image acquisition. Most remote sensing studies tend to rely on pre- and postoutbreak images to detect spectral response differences resulting from insect defoliation. In the case of the gypsy moth, an additional image taken during the refoliation period was used to increase evidence of the forest pest (Hurley et al., 2004). In general, the greater the number of images required, the greater the likelihood that remote sensing, particularly from satellite platforms, will fail due to the decreasing likelihood of finding cloud-free images. There are, however, 29 remote sensing satellites in orbit and 34 planned, of which 70% will offer a spatial resolution of 2 to 36 m with a range of spectral configurations (Stoney, 2004). As the number of these remote sensing satellite sensors increases, the likelihood of obtaining a cloud-free image during the narrow time periods when spectral changes are at its maximum will increase. The opportunities to acquire imagery ranging from high (e.g., submeter pixel size) to low spatial resolution (e.g., 1-km pixel size) are obviously increasing at an unprecedented rate that should help ensure that future image data will be available during the narrow time periods necessary to capture damage from pest defoliation.

This section has outlined the biology of these forest pests, their manifestation of damage, and how it influences the timing of remote sensing image acquisition. Logical questions that follow include: What has been the primary method used in defoliation surveys? Which remote sensing methods have been employed in mapping defoliation?

ASSESSMENT OF INSECT DEFOLIATION

AERIAL SURVEY TO ASSESS INSECT DEFOLIATION

Aerial sketch mapping is the process of a trained observer delineating damaged areas viewed from an aircraft onto a map. It is the most frequently used method to collect information on the location, intensity, and area affected by forest pests (Ciesla, 2000). These data are used in support of pest management and provincial and national reporting on the status of forest pests (e.g., Alberta Sustainable Resource Development, 2003; Hall et al., 1998; Simpson and Coy, 1999; USDA, 2004). Such surveys are quick and timely as they are undertaken when pest damage is most observable and are considered cost-effective because of the large areas that can be observed rapidly (Ciesla, 2000). Notably, the accuracy of such surveys has been questioned due to the subjectivity of observer ratings, the ability to detect defoliated areas, and the determination of the exact locations of observed defoliation on the map (Ciesla, 2000; MacLean and MacKinnon, 1996). Variable weather conditions affect visibility and aircraft stability, which can influence sketch map accuracy (MacLean and MacKinnon, 1996). Further, the damage label assigned to a delineated polygon is typically only an expression of the predominant level of damage since other levels of damage severity may occur within a polygon that has been mapped as moderate or severe. Aerial sketch map surveys are also impaired by the lack of time to record details, and the maps produced are not spatially precise because inclusions of large nondamaged areas and nonsusceptible land cover types result in overestimates of the actual area damaged (Harris and Dawson, 1979). In Canada, most forest health surveys conducted by provincial agencies tend to be focused on managed lands, and thus areas outside jurisdictional interest are not generally mapped for pest activity. Despite these compromising issues, aerial sketch map surveys remain in prevalent use.

Perhaps the motivators for the continual use of aerial sketch map surveys is that the method guarantees data acquisition, it maintains the continuity of long-term records on a national scale, and efforts are under way to improve the consistency and quality of conditions under which these surveys are undertaken. In particular, training and education (British Columbia Ministry of Forests, 2005), survey standards (USDA, 1999), and technological innovations such as digital capture systems based on global positioning system technology combined with geographic information systems and portable computers (Ciesla, 2000; USDA, 2003) are in use to improve the quality and consistency of these products. In particular, moving toward digital capture of sketch map surveys greatly improves the efficiency of map production and helps to reduce transfer errors because the delineated line work can be accomplished immediately instead of requiring manual digitization processes after the aerial survey flights are completed. Given these method refinements and how these survey results have been used in reporting, there is the likelihood that these surveys will continue in the near future. Aerial surveys provide information on the causal agent of forest damage and their approximate location. While inventory data are sometimes used to obtain more realistic areas of defoliation, this practice is not common. Remote sensing can complement this information by generating more spatially precise and detailed defoliation maps from which its impact on the forest resource could be determined. A review of the remote sensing methods that have been used for insect defoliation illustrates the degree that they have been successful in obtaining information of operational relevance (i.e., used by those in forest management).

REMOTE SENSING TO ASSESS INSECT DEFOLIATION

Selected remote sensing studies were compiled to identify trends in sensor, image data, image preprocessing, analysis method, and type of insect defoliation data that were used to assess the six insect defoliators covered in this chapter (Table 4.2). Studies ranged from 1984 to 2004, of which Landsat TM, Enhanced Thematic Mapper Plus (ETM+), and SPOT data were most frequently employed in insect defoliation studies. More than half of the studies employed two or more dates of image data, representing before- and after-defoliation time periods. Timing of image acquisition coincident with the period when the manifestation of damage is most visually obvious remains the most important criterion for image selection. Some

TABLE 4.2
Sample of Remote Sensing Studies Applied to the Six Insect Defoliators

Insect	Study area	Species	Sensor[1]	Image data: Single vs. multidate	Analysis method[2]	Damage classification: class or continuous	Reference
Aspen defoliators	Alberta	Trembling aspen	Landsat MSS	Multi-date: June 6/1977, June 8/1988	No AC[1]	Class data	Hall et al. 1984
	Alberta	Trembling aspen	Landsat TM	July 21, 1999, July 19, 2001	AC, modeling of changes in LAI	Continuous	Hall et al. 2003
Gypsy moth	Michigan	Oak	Landsat TM, SPOT	June 29, 1988 June 27, 1988	No AC, supervised and unsupervised	Class data	Joria et al. 1991
	Virginia	Oak	SPOT HRV-XS SPOT HRV-XS	June 15, 1987 July 4, 1988	Principle components, image differencing, spectral temporal change classn., Post-classn. change detection	Class data	Muchoney and Haack 1994
	Ohio	Oak	Landsat TM, ETM+ Landsat ETM+	Early June, Late June, Late July	AC, Infrared simple ratio and image differencing	Class data	Hurley et al. 2004
	Maryland Pennsylvania	Oak		Aug. 4, 1999, Aug. 22, 2000, July 24, 2001	AC, Tasselled Cap transformation, change vector analysis	Continuous via frass deposition	Townsend et al. 2004

Insect	Location	Host	Sensor[1]	Date	Method[2]	Data type	Reference
Spruce budworm	Newfoundland	Balsam fir	SPOT HRV MLA	Aug. 27, 1991	No AC, vegetation indices, discriminant function	Class data	Franklin and Raske 1994
	Quebec	Balsam fir White spruce	Landsat TM	July 22, 1986	AC, image segmentation	Class data	Chalifoux et al. 1998
Hemlock looper	Newfoundland	Balsam fir	SPOT HRV MLA	Aug. 29, 1987	No AC, supervised classification	Class data	Franklin 1989
	Newfoundland	Balsam fir	Landsat TM	Aug. 6, 1990	No AC, correlation and Discriminant analysis	Class data	Luther et al. 1991
	Quebec	Balsam fir, eastern hemlock	SPOT Vegetation	10-day composites: June 1-10 to 21-30	AC, multiple logistic regression	Class data	Fraser and Latifovic *in press*
Jack pine budworm	Saskatchewan	Jack pine	Landsat TM	July 20, 1984, Aug. 11, 1986, Aug. 30, 1987	AC, unsupervised classification	Class data	Hall et al. 1995
	Wisconsin	Jack pine	Landsat TM	June 21, 1984	No AC, supervised classification	Class data	Hopkins et al. 1988
	Northern Wisconsin	Jack pine	Landsat TM	June 14, 1987, May 10, 1992, Aug 1, 1993	AC, spectral mixture analysis	Continuous budworm population numbers	Radeloff et al. 1999

[1] Landsat MSS, Landsat Multispectral Scanner; Landsat TM, Landsat 4 or 5 Thematic Mapper; Landsat ETM+, Landsat 7 Enhanced Thematic Mapper Plus; SPOT, Satellite Pour l'Observation de la Terre; SPOT HRV-XS, SPOT High Resolution Visible, Multispectral; SPOT HRV MLA, SPOT High Resolution Visible, Multispectral Linear Array.

[2] AC, atmospheric correction procedures were employed.

studies of insect pests, such as spruce budworm and hemlock looper, that result in foliage color change were reported with single-date images. In terms of atmospheric preprocessing, only half of the studies employed atmospheric correction or image normalization procedures, but most of the more recent studies after 1998 employed such procedures. This observation may be reflective of the developing science by which change detection studies now recommend image preprocessing procedures such as radiometric and atmospheric correction, geometric rectification, and topographic normalization processes be performed prior to analysis (Lu et al., 2004).

Remote sensing methods employed in defoliation studies range from classification to modeling, and there appears to be no consistent approach for mapping (Table 4.2). Given the range of damage patterns, from physical loss of foliage to a change in foliage color, this finding was not unexpected. Image band ratios, transformations such as principle components and tasseled cap, image differencing, and various image classification approaches comprise the most frequently used approaches for mapping defoliation. Particularly promising results were reported from spectral mixture analysis, discriminant analysis, multiple logistic regression, and modeling changes in leaf area (Table 4.2). The type of information needed, however, is a driver for selection of method. Clearly, there are opportunities for developing more standardized procedures. To select the appropriate scale and image-processing method, the user requires a clear understanding of the problem and the information needed in relation to the biology and damage caused by the forest pest. Refer to Chapter 2, this volume, on scale and image-processing methods for additional details on image processing and spatial data modeling for capturing disturbances.

Only 3 of the 14 studies reviewed employed continuous estimates of insect defoliation damage (Table 4.2). Aerial sketch map surveys or ground ocular assessments are typically subjective, which results in relatively broad classes of defoliation damage that are frequently used in remote sensing studies. Defining the spectral basis for these broad damage limits will remain a problem until more continuous estimates of defoliation damage become available. To illustrate many of the fundamental concepts and trends summarized in this chapter, the following case study shows the integration of biological knowledge with remote sensing methods to map and monitor aspen defoliation.

CASE STUDY: REMOTE SENSING OF ASPEN DEFOLIATION

INTRODUCTION

Trembling aspen is the most widely distributed North American tree species (Perala, 1990) and, from ecological and commercial perspectives, the most important deciduous tree species in the North American boreal forest (Hogg et al., 2002; E. B. Peterson and Peterson, 1992). Its geographic distribution extends from Atlantic Canada to the Pacific coast of Alaska and as far south as Mexico (E. B. Peterson and Peterson, 1992). Repeated defoliation by insects, in combination with drought, severe early spring freeze-thaw events, and fungal pathogens has caused reduced growth and dieback of aspen in different parts of North America (Brandt et al., 2003; Hogg et al., 2002).

In northern Alberta, Canada, an outbreak of large aspen tortrix defoliation was recorded by aerial sketch map surveys during the late 1990s and early 2000s, followed by collapse in 2004. This case study demonstrates how a four-year (2001 — 2004) time series of Landsat TM images can be used to map trembling aspen defoliation patterns by monitoring changes in LAI at the approximate same time of year through comparison with the preoutbreak year (1999). LAI is considered a fundamental biophysical measure of the forest canopy due to its governing role in many ecophysiological processes, such as canopy light interception, evapotranspiration, and photosynthesis (Running et al., 1989). In turn, defoliation and dieback influence the tree's vegetative biomass, which can be monitored by satellite mapping of changes in leaf area.

METHODS: HOW IS ANNUAL MONITORING OF DEFOLIATION PATTERNS ACHIEVED?

Study Area

The study area was located near the town of High Level in north central Alberta situated at 58.5°N, 116.2°W. The species occurring in the study area depended on moisture and site conditions and were characterized by pure and mixed stands of trembling aspen, balsam poplar, and white spruce. The study area was selected in part because of reported outbreaks of large aspen tortrix, which had been severely defoliating many trembling aspen stands (R. J. Hall et al., 2003). The area was also part of a long-term research project called CIPHA (Climate Change Impacts on Productivity and Health of Aspen) that was initiated in 2000 to monitor changes in aspen health as a result of changes in climate in combination with insect defoliation (Hogg et al., 2005).

Aerial and Field Data Collection

Aerial sketch maps are produced annually by an experienced provincial forest health surveyor, who delineates polygons onto a 1:250,000 topographic map while flying in a fixed-wing Cessna 210 aircraft at an elevation range of 900 to 1200 m above sea level at a speed of 270 to 300 km/hr (Maximchuk, 2005). These maps were obtained from the province for the 1999, 2001, 2002, 2003, and 2004 survey years. Each manually sketched polygon was assigned a defoliation severity rating of nil-light (<35%), moderate (35–70%), or severe (>70%) as defined in the Alberta Land and Forest Service Aerial Survey Manual (Alberta Sustainable Resource Development, 2002). Larger areas of defoliation may include a combination of light, moderate, and severe defoliation but are labeled by the most frequently occurring or dominant severity rating for that area. Since aerial sketch maps generally provide an overview of defoliation extent and severity, they were useful in identifying where to locate field plots and to provide a general level of validation over the landscape.

Field plots were located within the strata defined by the aerial sketch map survey (Figure 4.2, Method 1a). Estimates of defoliation were based on mean defoliation ratings of 10 to 20 randomly selected trees per plot during the peak defoliation period of 22–25 June 2001. Defoliation of individual trees was visually estimated with the aid of binoculars to 10% classes (Michaelian et al., 2001) (Figure 4.2, Method 1b). Ten optical LAI-2000 measurements were taken in each selected plot (Figure 4.2, Method 2) and averaged for validating the estimation of LAI on the 2001 image. Given that it was not feasible to measure LAI in the field for each year of the time series, an indicator of how well LAI could be estimated on the image was derived through validating results from one of the image years. The optical LAI-2000 measurements were also validated by comparison with up to ten litter traps that were established in each sample plot (R. J. Hall et al., 2003). The geographic location of each plot was recorded with the aid of a Trimble Pro-XR global positioning satellite receiver, which had data subsequently differentially corrected to within 5 m of its true location with local base station data.

Satellite Remotely Sensed Data

Landsat TM and ETM+ image data (Table 4.3) were acquired to map the peak occurrence of aspen defoliation from 2001 to 2004, and a 1999 image was used as the predefoliation image (Figure 4.2, Method 3). The image selection criteria were defined by (a) an image acquisition date of early June to early July to capture the outbreak stage of defoliation prior to refoliation of trembling aspen that tends to occur during later July (Table 4.1); (b) a nondefoliated image to associate changes in spectral response from healthy deciduous stand conditions to defoliated stand conditions; and (c) a relatively cloud-free overlap region between the images. Relating the period of peak defoliation at a particular stage of vegetative phenology to the timing of image acquisition is the most important criterion when selecting images. While no cloud-free Landsat image was available from mid- to late June over the area of interest, there was a relatively consistent time series available in early July that was deemed acceptable (Table 4.3).

The Landsat images were processed to retrieve top-of-atmosphere directional reflectance (Figure 4.2, Method 3a). The Landsat data corresponded to Level 1G at sensor radiance systematic corrected data that was processed using the PGS processor by Radarsat International. Radiometric calibration was performed using information provided in the image header files. An iterative dense dark vegetation atmospheric correction approach was applied to all scenes in this study using information extracted from selected dense dark vegetation pixels across various spectral wavelengths to determine the contribution of aerosols influencing the pixel radiances recorded at the satellite sensor. The 6S radiative transfer code (Vermote et al., 1997) was used iteratively to determine an estimate of surface reflectance given top-of-atmosphere reflectance and an estimate of aerosol optical depth. This approach provided a systematic and repeatable method for both radiometric calibration and atmospheric correction of multitemporal Landsat imagery as these factors are essential to produce precise estimates of LAI based on the broadband spectral vegetation indices that were used (Fernandes et al., 2003). All the images

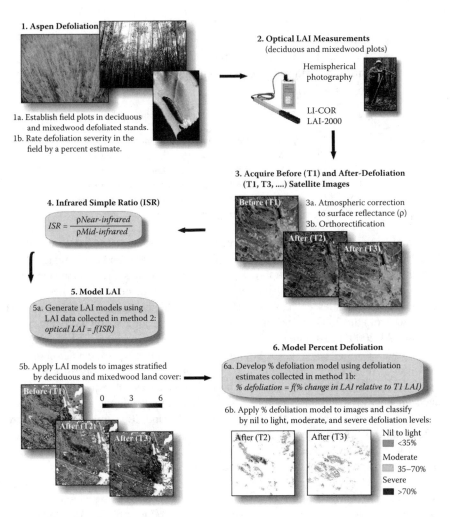

FIGURE 4.2 (See color insert following page 146.) Method flowchart to generate a time series of aspen defoliation maps based on changes in LAI recorded between pre- and postdefoliation satellite images.

were then georeferenced to the Lambert conformal conic projection using a nearest-neighbor, first-order transformation (Figure 4.2, Method 3b).

DETERMINING LEAF AREA INDEX FOR BEFORE AND AFTER DEFOLIATION TIME PERIODS

To determine the change in leaf area attributable to defoliation, a model that relates an image ratio or vegetation index to leaf area was needed for each image date so that differences resulting from defoliation could be computed. Studies suggested that the infrared simple ratio (ISR) (computed as Landsat ETM+ band 4/Landsat ETM+ band 5) was a more robust indicator of LAI (Fernandes et al., 2003) than

TABLE 4.3
Landsat TM and ETM+ Image Data for the Case Study for Mapping Aspen Defoliation Induced by Insect Activity

Path	Row	Sensor	Year	Day	Aspen defoliator
46	19	ETM+	1999	July 21	Noninfestation year
46	19	ETM+	2001	July 10	Large aspen tortrix
47	19	TM	2002	July 12	Large aspen tortrix
47	19	TM	2003	July 15	Large aspen tortrix
46	19	TM	2004	July 10	Forest tent caterpillar

the simple ratio or reduced simple ratio reported in earlier studies (Chen et al., 2002). The ISR was computed for each of the 2001, 2002, 2003, and 2004 image dates (Figure 4.2, Method 4). Structural regressions between Landsat reflectance measurements based on the ISR and in situ optical LAI measurements were applied to each image date (Fernandes et al., 2003) (Figure 4.2, Method 5a). Validation of LAI estimates from the 2001 image was undertaken through comparison with field-based optical LAI-2000 measurements (Figure 4.2, Method 2). Model results were within a maximum absolute error of 1 LAI unit with a root-mean-square deviation of 0.37 LAI unit, a remarkable result considering it was originally developed for national applications (R. J. Hall et al., 2003). The ISR model was subsequently applied to each image date to produce an LAI image within the deciduous and mixed wood land cover types for each of the 2001, 2002, 2003, and 2004 image dates (Figure 4.2, Method 5b). Classified land cover for this region from the Earth Observation for Sustainable Development of forests program (Wulder et al., 2003) in collaboration with the Alberta Ground Cover Characterization project was used to stratify the images into deciduous and mixed wood cover types to ensure regions designated as coniferous, shrub, agriculture, or other nonsusceptible species types would not be modeled. The overall accuracy of land cover maps produced by the Alberta Ground Cover Characterization project is targeted at 80%.

SATELLITE MAPPING OF ASPEN DEFOLIATION

Because changes in LAI attributable to defoliation should be relative to the amount of LAI present before defoliation, the percentage change in LAI from 2001 to 2004 was computed relative to LAI in 1999. A model to estimate percentage defoliation as a function of change in LAI was derived based on percentage defoliation values measured in the field (Figure 4.2, Method 6a). This model was an exponential function with an R^2 of 0.77 (R. J. Hall et al., 2003). While this model was originally generated for descriptive purposes, additional field and image data could be used in an operational program to define better the relationship between percentage defoliation and percentage change in LAI relative to the before-defoliation year. This percentage defoliation model was subsequently applied to the defoliation images for

2001, 2002, 2003, and 2004. A median filter was employed on the resulting defoliation raster maps to remove isolated single pixels and to preserve homogeneous areas of pixels that more likely represent areas of aspen defoliation. A visual assessment of the image models within the aerial sketch-mapped areas suggested that most of these isolated pixels were located outside the delineated defoliated areas. The image models were then classified into the three broad defoliation severity classes (<35%, 35–70%, and >70%) to produce a thematic map depicting the patterns of defoliation severity for each image year.

RESULTS: REFINEMENT OF SKETCH MAP SURVEY AREA

The aerial sketch map surveys provide a broad indicator of the annual areal extent of aspen defoliation. These surveys overestimate the actual area of damage because they include areas not susceptible to aspen defoliation, such as conifer forest, agricultural lands, and water bodies. Within the study area selected for this case study, the deciduous and mixed wood land covers were approximately 232,000 Ha and 302,000 Ha, respectively, which amounted to an approximate 530,000 Ha of forested land that was potentially susceptible to aspen defoliation. The total defoliated area derived from aerial sketch mapping was nil in 1999, but the area increased to more than 350,000 Ha annually between 2001 and 2003, followed by a reduction to 90,000 Ha in 2004 due to the collapse of the large aspen tortrix outbreak. Intersecting the deciduous and mixed wood land cover with the aerial sketch maps excluded areas not subject to defoliation. The resulting area of susceptible forest was approximately 70,000 Ha for deciduous forest and 95,000 Ha for mixed wood forest from 2001 to 2003. The susceptible area obtained in 2004 was approximately 19,000 Ha for deciduous forest and 37,000 Ha for mixed wood forest.

Once the amount of susceptible area was determined, it was simpler to isolate the defoliated forest within it. Translating the percentage reduction in leaf area from the satellite images to percentage defoliation resulted in a much smaller area of defoliation within each of the deciduous and mixed wood land cover types (Figure 4.3). Aspen defoliation in the deciduous land cover was approximately 20,000 to 30,000 Ha from 2001 to 2003 and less than 10,000 Ha in 2004 (Figure 4.3a). The yearly pattern of defoliation was markedly different between the deciduous (Figure 4.3a) and mixed wood (Figure 4.3b) land cover types, which was attributed to increased complexity of the mixed wood stands due to the conifer component. The large decrease in defoliated area observed in 2004 was consistent with the reported collapse of defoliation from the aerial sketch map surveys. While the accuracy of the smaller areas is difficult to validate, we can at least compare the general trends reported from the aerial sketch map survey with that derived from the remote sensing mapping of defoliation.

There was a similar trend in the patterns of defoliation mapped from the remote sensing time series compared to that mapped from the aerial surveys (Figure 4.4). The similarity in these yearly trends is a strong indicator that the remote sensing approach does provide a reasonable representation of the actual trends as represented in the aerial sketch map data over the four-year time period. The key difference in these results is that the aerial survey sketch maps represent the total areal extent of

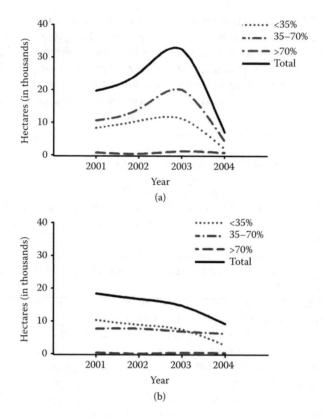

FIGURE 4.3 Area (hectares) impacted by light to nil (<35%), moderate (35–70%), and severe (>70%) defoliation for (a) deciduous and (b) mixed wood stands.

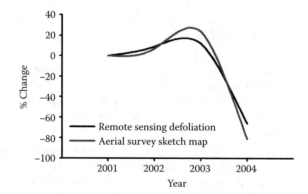

FIGURE 4.4 Percentage change of total defoliation mapped from 2001 for remote sensing defoliation and aerial survey sketch maps.

defoliation, whereas the remote sensing derivation is closer to representing the net areal extent that appears to have sustained defoliation.

A compilation of the remote sensing defoliation maps from 2001 to 2004 for a subsection of the study area is illustrated in Figure 4.5. The sketch maps and image composites provide a visual comparison of the results derived from aerial survey and remote sensing. The image enhancement distinguishes areas of aspen defoliation by a grayish color tone, with healthy deciduous and mixed wood stands depicted by light and dark orange tones, respectively. By comparing the defoliated areas on the images to the defoliation sketch maps, it was obvious where the remote sensing estimates more precisely defined the location of defoliation activities. The sketch maps, however, do provide a useful framework for associating the remote sensing patterns to the general areas of insect infestation. While the remote sensing changes in leaf area did identify areas of greatest change, without ancillary information it would be difficult to identify the cause of the change; an integration of both methods therefore appears best. The aerial sketch map would identify the causal agent and the general region where this activity is taking place, and remote sensing of disturbance caused by defoliation (Figure 4.2) would provide the mechanism for more precise mapping of the actual areas of defoliation.

SUMMARY

One of the major natural disturbances on forest landscapes is caused by insect defoliators. While aerial and ground surveys are the means by which these disturbances are typically recorded, there has been considerable interest to explore the advantages of using remote sensing to help meet the information needs of government and industry. These advantages include achieving more consistent and precise mapping of pest activities for reporting and assessing impacts on sustainability of forest ecosystems.

A conceptual model for more successful use of remote sensing was proposed through the integration of pest, host, and remote sensing knowledge. A table of comparative biology and damage patterns for six major defoliators summarized information about the manifestation of damage and the timing for which these activities occur relative to host tree phenology. This information is required to define the spectral basis for pest damage and timing of image acquisition relative to the causal agent. A review of remote sensing studies undertaken specifically to detect and map defoliation from the six pests revealed that a wide range of data and methods of analysis have been employed. A case study example with a comprehensive methods flowchart was assembled to illustrate the model application for mapping and monitoring of aspen defoliation over a four-year time series.

A challenge is how to move forward toward an operational defoliation mapping program that uses remote sensing. While aerial surveys are subjective and not spatially precise, sketch maps do provide a strong advantage in that data acquisition is guaranteed. An approach to improve on these surveys is to develop an integrated program since the use of remote sensing alone for detection is more difficult than its use for mapping. Due to the large regions over which insect pests may occur, the selection of remotely sensed imagery is best undertaken within the framework of

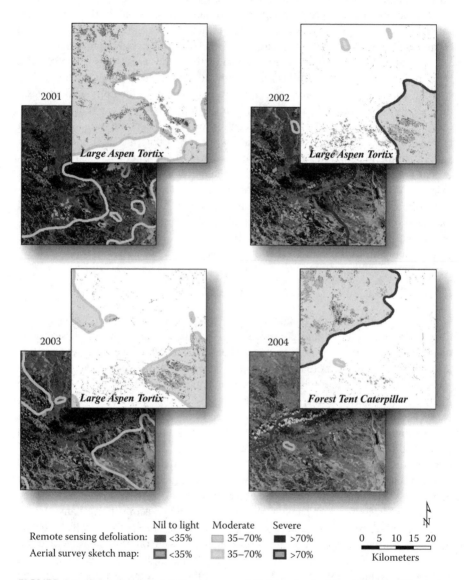

FIGURE 4.5 (See color insert following page 146.) A subset of the study area illustrating the pattern of insect activity from 2001 to 2004. The aspen defoliation mapped by remote sensing provides a more realistic estimate of the severity and extent of forest damage than the aerial survey sketch maps. Image composites consist of Landsat band 4 as red, band 5 as blue, and band 3 as green using an adaptive stretch enhancement.

aerial surveys. These surveys provide information on the likely causal agent and the general location of pest defoliators to focus the image selection and acquisition activities. Subsequent analysis of remotely sensed images provides more precise mapping of the location and severity of the insect defoliation event.

ACKNOWLEDGMENTS

Processing of the Landsat images in the case study to surface reflectance by C. Butson, of the Canadian Forest Service, is greatly appreciated. We are grateful to R. Lay, who assisted in the search and organization of the literature. Jim Ellenwood, Program Manager of Remote Sensing and Image Analysis for the Forest Health Technology Enterprise Team, United States Department of Agriculture — Forest Service, provided linkages to information sources. Review of the insect biology section by Dr. J. Volney of the Canadian Forest Service of Natural Resources Canada is appreciated.

REFERENCES

Alberta Sustainable Resource Development. (2002). *Forest Health Aerial Survey Summary.* Alberta Sustainable Resource Development, Forest Management Branch, Land and Forest Division, Edmonton, AB, Canada. Retrieved May 2004 from http://www3.gov.ab.ca/srd/forests/health/data_standards.html.

Alberta Sustainable Resource Development. (2003). *2003 Annual Report: Forest Health in Alberta.* Department of Sustainable Resource Development, Public Lands and Forests Division, Forest Management Branch, Forest Health Section, Edmonton, AB, Canada.

Alfaro, R.I. (1988). Pest damage in forestry and its assessment. *Northwest Environ. J., 4,* 279–300.

Brandt, J.P., Cerezke, H.F., Mallett, K.I., Volney, W.J.A., & Weber, J.D. (2003). Factors affecting trembling aspen (*Populus tremuloides* Michx.) health in the boreal forest of Alberta, Saskatchewan, and Manitoba, Canada. *Forest Ecology and Management, 178*, 287–300.

British Columbia Ministry of Forests. (2005). *Aerial Overview Surveys Training Program.* British Columbia Ministry of Forests, Forest Practices Branch. Retrieved April 21, 2005, from http://www.for.gov.bc.ca/pscripts/hfd/mtc/course.asp?courseid=32.

Cadogan, B.L. (1995). Jack pine budworm, *Choristoneura pinus.* In J.A. Armstrong and W.G.H. Ives (Eds.), *Forest Insect Pests in Canada* (pp.123–126). Natural Resources Canada, Canadian Forest Service, Ottawa, ON, Canada.

Cerezke, H.F. and Volney, W.J.A. (1995). *Forest insect pests in the Northwest region.* In Forest insects pests in Canada. Edited by J.A. Armstrong and W.G.H. Ives. Science and Sustainable Development Directorate, Canadian Forest Service, Natural Resources Canada, Ottawa, Ontario. pp. 59–72.

Chalifoux, S., Cavayas, F., and Gray, J.T. (1998). Map-guided approach for the automatic detection on Landsat TM images of forest stands damaged spruce budworm. *Photogrammetric Engineering and Remote Sensing, 64*, 629–635.

Chen, J.M., Pavlic, G., Brown, L., Cihlar, J., Leblanc, S.G., White, H.P., Hall, R.J., Peddle, D., King, D.J., Trofymow, J.A., Swift, E., Van der Sanden, J., and Pellikka. P. (2002). Validation of Canada-wide leaf area index maps using ground measurements and high and moderate resolution satellite imagery. *Remote Sensing of Environment, 80*, 165–184.

Ciesla, W.M. (2000). *Remote Sensing in Forest Health Protection*. USDA Forest Service, FHTET Rep. 00-03. 266p.

Coulson, R.N. and Witter, R.A. (1984). *Forest Entomology: Ecology and Management*. John Wiley and Sons, Toronto, ON, Canada. 669p.

Dottavio, C.L. and Williams, D.L. (1983). Satellite technology: an improved means for monitoring forest insect defoliation. *Journal of Forestry, 81*, 30–35.

Fellin, D.G., and Dewey, J.E. (1982). *Western spruce budworm*. Forest Service, States Department of Agriculture, Forest Insect and Disease Leaflet, 53, 10p.

Fernandes, R.A., Butson, C.B., Leblanc, S.G., and Latifovic, R. (2003). Landsat-5 TM and Landsat-7 ETM+ based accuracy assessment of leaf area index products for Canada derived from SPOT-4 Vegetation data. *Canadian Journal of Remote Sensing, 29*, 241–258.

Franklin, S.E. (1989). Classification of hemlock looper defoliation using SPOT HRV imagery. *Canadian Journal of Remote Sensing, 15*, 178–182.

Franklin, S.E. (2001). *Remote Sensing for Sustainable Forest Management*. Lewis Publishers, Boca Raton, FL. 407p.

Franklin, S.E. and Raske, A.G. (1994). Satellite remote sensing of spruce budworm forest defoliation in western Newfoundland. *Canadian Journal of Remote Sensing, 20*, 37–48.

Fraser, R.H. and Latifovic, R. (2005). Mapping insect-induced tree defoliation and mortality using coarse spatial resolution satellite imagery. *International Journal of Remote Sensing, 26*, 193–200.

Furniss, R.L. and Carolin, V.M. (1977). *Western Forest Insects*. USDA For. Serv. Misc. Publ. No. 1339.

Hall, P.J., Bowers, W.W., and Hirvonen, H. (1998). *Forest Insect and Disease Conditions in Canada*. Natural Resources Canada, Canadian Forest Service, Ottawa, ON, Canada. 72p.

Hall, P.J. and Moody, B.H. (1994). *Forest Depletions Caused by Insects and Diseases in Canada, 1982 — 1987*. Natural Resources Canada, Canadian Forest Service, Ottawa, ON, Canada. Information Report ST-X-8. 14p.

Hall, R.J., Crown, P.H., and Titus, S.J. (1984). Change detection methodology for aspen defoliation with Landsat MSS digital data. *Canadian Journal of Remote Sensing, 10*, 135–142.

Hall, R.J., Crown, P.H., Titus, S.J., and Volney, W.J.A. (1995). Evaluation of Landsat Thematic Mapper data for mapping top kill caused by jack pine budworm defoliation. *Canadian Journal of Remote Sensing, 21*, 388–399.

Hall, R.J., Fernandes, R.A., Butson, C., Hogg, E.H., Brandt, J.P., Case, B.S., and Leblanc, S.G. (2003). Relating aspen defoliation with changes in leaf area from field and satellite remote sensing perspectives. *Canadian Journal of Remote Sensing, 2*, 299–313.

Hall, R.J., Still, G.N., and Crown, P.H. (1983). Mapping the distribution of aspen defoliation using LANDSAT color composites. *Canadian Journal of Remote Sensing, 9*, 86–91.

Hall, R.J., Volney, W.J.A., and Wang, Y. (1998). Using GIS to associate forest stand characteristics with top kill resulting from defoliation by the jack pine budworm. *Canadian Journal of Forest Research, 28*, 1317–1327.

Harris, J.W.E. and Dawson, A.F. (1979). *Evaluation of Aerial Forest Pest Damage Survey Techniques in British Columbia*. Canadian Forest Service, Pacific Forestry Centre, Victoria, BC, Canada. Information Report BC-X-198. 22p.

Heikkilä, J., Nevalainen, S., and Tokola, T. (2002). Estimating defoliation in boreal coniferous forests by combining Landsat TM, aerial photographs and field data. *Forest Ecology and Management, 158*, 9–23.

Herms, D.A. and Shetlar, D.J. (2002). Assessing options for managing gypsy moth. *The Ohio State University Extension Factsheet, HYG-2175–02*. Retrieved April 10, 2005, from http://ohioline.osu.edu/hyg-fact/2000/pdf/2175.pdf.

Hogg, E.H., Brandt, J.P., and Kochtubajda, B. (2002). Growth and dieback of aspen forests in northwestern Alberta, Canada, in relation to climate and insects. *Canadian Journal of Forest Research, 32*, 823–832.

Hogg, E.H., Brandt, J.P., and Kochtubajda, B. (2005). Factors affecting interannual variation in growth of western Canadian aspen forests during 1951 — 2000. *Canadian Journal of Forest Research, 35*, 610–622.

Hopins, P.F., MacLean, A.L., and Lillesand, T.M. (1988). Assessment of Thematic Mapper imagery for forestry applications under Lake States conditions. *Photogrammetric Engineering and Remote Sensing, 54*, 61–68.

Howse, G.M. (1984). Insect pests of jack pine: biology, damage and control. *Proceedings Jack Pine Symposium* (pp. 131–138). Environment Canada, Canadian Forest Service, Great Lakes Forestry Research Centre, Sault Ste. Marie, ON, Canada. Canada-Ontario Joint Forest Research. Committee Proceedings 0-P-12.

Hurley, A., Watts, D., Burke, B., and Richards, C. (2004). Identifying gypsy moth defoliation in Ohio using Landsat data. *Environmental and Engineering Geoscience, 10*, 321–328.

Ives, W.G.H. and Wong, H.R. (1988*). Tree and shrub insects of the prairie provinces*. Canadian Forest Service, Northern Forestry Center, Edmonton, Alberta, Canada. Information Report NOR-X-292. 327p.

Jobin, L. (1995). Gypsy moth, *Lymantria dispar*. Forest insect pests in Canada. Edited by Armstrong, J.A., and W.G.H. Ives. Natural Resources Canada, Can. For. Serv., Science and Sustainable Development Directorate, Ottawa, Canada. pp. 134–139.

Joria, P., Ahearn, S., and Connor, M. (1991). A comparison of the SPOT HRV and Landsat Thematic Mapper for detecting gypsy moth defoliation in Michigan. *Photogrammetric Engineering and Remote Sensing, 57*, 1605-1612.

Kulman, H.M. (1976). Effects of insect defoliation on growth and mortality of trees. *Annual Review of Entomology, 16*, 289–324.

Kulman, H.M., Hodson, A.C., and Duncan, D.P. (1963). Distribution and effects of jack pine budworm defoliation. *Forest Science, 9*, 146–156.

Leatherman, D.A., Farmer, D.S., and Hill, D.S. (1995). *Gypsy Moth*. Colorado State University, Fort Collins, CO. Trees and Shrubs, Cooperative Extension Newsletter, Number 5.539.

Leckie, D.G., Cloney, E., and Joyce, S.P. (2005). Automated detection and mapping of crown discolouration caused by jack pine budworm with 2.5-m resolution multispectral imagery. *International Journal of Applied Earth Observations and Geoinformation, 7*, 61–77.

Leckie, D.G. and Ostaff, D.G. (1988). Classification of airborne multispectral scanner data for mapping current defoliation caused by the spruce budworm. *Forest Science, 34*, 259–275.

Leckie, D.G., Teillet, P.M., Fedosejevs, G., and Ostaff, D.P. (1988). Reflectance characteristics of cumulative defoliation of balsam fir. *Canadian Journal of Forest Research, 18*, 1008–1016.

Lu, D., Mausel, P., Brondizio, E., and Moran, E. (2004). Change detection techniques. *International Journal of Remote Sensing, 25*, 2365–2407.

Luther, J.E., Franklin S.E., and Hudak, J. (1991). Satellite remote sensing of current-year defoliation by forest pests in western Newfoundland. In S.E. Franklin, M.D. Thompson, and F.J. Ahern (Eds.), *Proceedings of the 14th Canadian Symposium on Remote Sensing* (pp. 192–198) May 6–10, 1991, Calgary, Alberta. Canadian Remote Sensing Society, Ottawa, ON, Canada.

Luther, J.E., Franklin, S.E., Hudak, J., and Meades, J.P. (1997). Forecasting the susceptibility and vulnerability of balsam fir stands to insect defoliation with Landsat Thematic Mapper data. *Remote Sensing of Environment, 59*, 77–91.

MacLean, D.A. (1990). Impact of forest pests and fire on stand growth and timber yield: implications for forest management planning. *Canadian Journal of Forest Research, 20*, 391–404.

MacLean, D.A. and Ebert, P. (1999). The impact of hemlock looper [*Lambdina fiscellaria fiscellaria* (Guen.)] on balsam fir and spruce in New Brunswick, Canada. *Forest Ecology and Management, 120*, 77–87.

MacLean, D.A. and MacKinnon, W. (1996). Accuracy of aerial sketch-mapping estimates of defoliation. *Canadian Journal of Forest Research, 26*, 2099–2108.

Martineau, R. (1984). *Insects harmful to forest trees.* Multiscience Pub. Ltd., Montreal, Que. 261pp.

Maximchuk, M. (2005). Personal communications. September 1.

Michaelian, M., Brandt, J.P., and Hogg, E.H. (2001). Climate Change Impacts on the Productivity and Health of Aspen (CIPHA): Methods Manual. Natural Resources Canada, Canadian Forest Service, Northern Forestry Centre, Edmonton, AB, Canada. Unpublished report. 42p.

Moody, B.H. (1986). *The Jack Pine Budworm History Of Outbreaks, Damage and FIDS Sampling and Prediction System in the Prairie Provinces. Jack Pine Budworm Information Exchange* (pp. 15–22). Manitoba Department of Natural Resources, Winnipeg, MB, Canada.

Muchoney, D.M. and Haack, B.N. (1994). Change detection for monitoring forest defoliation. *Photogrammetric Engineering and Remote Sensing, 60*, 1243–1251.

Murtha, P.A. (1982). Detection and analysis of vegetation stresses. *Remote Sensing for Resource Management* (pp. 141–158). Soil Conservation Society of America, Ankeny, IA.

Nealis, V.G. and Erb, S. (1993). *A sourcebook for management of the gypsy moth.* Forestry Canada, Great Lakes Forestry Centre, Sault Ste. Marie, Ont. 48pp.

Ostaff, D.P. and MacLean, D.P. (1989). Spruce budworm populations, defoliation and changes in stand condition during an uncontrolled spruce budworm outbreak on Cape Breton Island, Nova Scotia. *Canadian Journal of Forest Research, 19*, 1077–1086.

Perala, D.A. (1990). *Populus tremuloides* Michx. — quaking aspen. In R.M. Burns and B.H. Honkala (Eds.), *Silvics of North America. Vol. 2. Hardwoods* (pp. 555–569). U.S. Department of Agriculture Handbook. Rand Corporation, Santa Monica, CA. 654p.

Peterson, D.J., Resetar, S., Brower, J., and Diver, R. (1999). *Forest Monitoring and Remote Sensing: A Survey of Accomplishments and Opportunities for the Future.* RAND Science and Technology Policy Institute, Washington, DC. Report MR-111.0-OSTP.

Peterson, E.B. and. Peterson, N.M. (1992). *Ecology, Management, and Use of Aspen and Balsam Poplar in the Prairie Provinces.* Forestry Canada, Northwest Region, Northern Forestry Centre, Edmonton, AB, Canada. Special Report 1. 252p.

Radeloff, V.C., Mladenoff, D.J., and Boyce, M.S. (1999). Detecting jack pine budworm defoliation using spectral mixture analysis: separating effects from determinants. *Remote Sensing of Environment, 69*, 156–169.

Raske, A.G., West, R.J., and Retnakaran, A. (1995). Hemlock looper, *Lambdina fiscellaria.* In J.A. Armstrong and W.G.H. Ives (Eds.), *Forest Insect Pests in Canada* (pp. 141–147). Natural Resources Canada, Canadian Forest Service, Ottawa, ON, Canada.

Riley, J.R. (1989). Remote sensing in entomology. *Annual Review of Entomology, 34*, 247–271.

Royle, D.D. and Lathrop, R.G. (1997). Monitoring hemlock forest health in New Jersey using Landsat TM data and change detection techniques. *Forest Science, 43*, 327–335.

Running, S.W., Nemani, R.R., Peterson, D.L., Band, L.E., Potts, D.F., Pierce, L.L., and Spanner, M.A. (1989). Mapping regional forest evapotranspiration and photosynthesis by coupling satellite data with ecosystem simulation. *Ecology, 70,* 1090–1101.

Shepherd, R.F., Gray, T.G., and Harvey, G.T. (1995). Geographical distribution of *Choristoneura* species (Lepitoptera: Tortricidae) feeding on *abies, picea,* and *pseudotsuga* in western Canada and Alaska. *Canadian Entomologist,* 127, 813–830.

Simpson, R. and Coy, D. (1999). *An Ecological Atlas of Forest Insect Defoliation in Canada 1980–1996. Natural Resources Canada.* Canadian Forest Service, Atlantic Forestry Centre, Fredericton, NB, Canada. Information Report M-X-206E. 15p.

Stoney, W.E. (2004). *ASPRS Guide to Land Imaging Satellites.* Mitretek Systems. Retrieved April 22, 2005, from http://www.asprs.org/news/satellites/satellites.html.

Townsend, P.A., Eshleman, K.N., and Welcker, C. (2004). Remote sensing of gypsy moth defoliation to assess variations in stream nitrogen concentrations. *Ecological Applications, 14,* 504–516.

United States Department of Agriculture. (1999). *Aerial Survey Standards.* United States Department of Agriculture–Forest Service in conjunction with State and Private Forestry Forest Health Protection. Retrieved April 20, 2005, from http://www.fs.fed.us/foresthealth/publications/id/standards_1099.pdf. 7p.

United States Department of Agriculture. (2003). *Aerial Survey GIS Handbook: Sketchmaps to Digital Geographic Information.* United States Department of Agriculture–Forest Service in conjunction with State and Private Forestry Forest Health Protection. Retrieved April 20, 2005, from http://www.fs.fed.us/foresthealth/publications/id/gis-handbook.pdf. 74p.

United States Department of Agriculture. (2004). *Forest Insect and Disease Conditions in the United States 2003.* United States Department of Agriculture, Forest Service, Forest Health Protection, Washington, DC. 142p.

Vermote, E.F., Tanré, D., Deuzé, J.L., Herman, M., and Morcrette, J.J. (1997). Second simulation of the satellite signal in the solar spectrum: an overview. *IEEE Transactions on Geoscience and Remote Sensing, 35,* 675–686.

Volney, W.J.A. (1985). Comparative population biologies of North American spruce budworms. In C.J. Sanders, R.W. Stark, E.J. Mullins, and J. Murphy (Eds.), *Recent Advances in Spruce Budworms Research. Proceedings of the CANUSA Spruce Budworms Research Symposium* (pp. 71–84). Canadian Forestry Service and United States Department of Agriculture–Forest Service, Ottawa, ON, Canada.

Volney, W.J.A. (1988). Analysis of historic jack pine budworm outbreaks in the Prairie Provinces of Canada. *Canadian Journal of Forest Research, 1,* 1152–1158.

Volney, W.J.A. and Fleming, R.A. (2000). Climate change and impacts of boreal forest insects. *Agriculture, Ecosystems and Environment, 82,* 283–294.

Volney, W.J.A. and McCullough, D.G. (1994). Jack pine budworm population behaviour in northwestern Wisconsin. *Canadian Journal of Forest Research, 24,* 502–510.

Wulder, M.A., Dechka, J., Gillis, M., Luther, J., Hall, R.J., Beaudoin, A., and Franklin, S.E. (2003). Operational mapping of the land cover of the forested area of Canada with Landsat data: EOSD land cover program. *The Forestry Chronicle, 79,* 1075–1083.

5 Using Remote Sensing to Map and Monitor Fire Damage in Forest Ecosystems

Jess Clark and Thomas Bobbe

CONTENTS

INTRODUCTION

Each year, wildfires burn millions of hectares all over the world, with estimates suggesting that fires annually burn an area half the size of China (The Nature

Conservancy, 2004). Although widely characterized as detrimental to the environment, fire can be a critical ecosystem component. For example, frequent low-intensity fires consumed much of the understory and downed woody debris, allowing old-growth forests to flourish. Ecologists theorize that some species adapted to fire, using it as a mechanism for seeding and further distribution. In Australia, one example is the eucalyptus tree, easily the most dominant species on that continent. The eucalyptus trees "amalgamated hundreds of species which were ideally predisposed to survive in an environment of increasing fire" (Pyne, 1991, p. 27). In some cases, this adaptation to fire was essential, especially in Australia, where fire is commonplace. "Fire had touched other continents, but it branded Australia" (Pyne, 1991, p. 66).

Early settlers in Australia used fire as a hunting tool, hunting along the fire front as frightened animals were flushed out of their habitats. European explorers reported that the Aborigine peoples in Australia used fire so often they would walk around with "fire sticks" (Pyne, 1991, p. 85) to create fire whenever they needed, whether for meals, hunting, or warmth. These human-caused fires facilitated a fire-adapted environment dependent on frequent low-intensity fires that maintained healthy vegetation conditions. When European settlers came to Australia and displaced Aborigines and their frequent fires, they reported rapid modification of vegetation structure and composition. Instead of grasses they had originally likened to English parks, they cited a thickening of scrub, enough eventually to close off open forests. The management practice of excluding fire from ecosystems where possible created significant fuel loads throughout Australia, therefore increasing the risk of catastrophic fire. "By the 1890s the consequences of removing Aboriginal fire became inescapable. By the 1980s, they required restitution" (Pyne, 1991, p. 133). Prescribed fire is now a frequently used practice in Australia.

Several other examples throughout the world have been noted regarding the effects of the removal of fire from fire-adapted ecosystems. In North America, the landscape was accustomed to frequent fire, nearly all naturally caused. Repeated fires created open forests and cleared much of the dense understory. With the arrival of Anglo-European settlers, natural fires became a hindrance. Fire suppression in the United States began in the early 1900s after large wildfires destroyed millions of hectares of forest, burned many homes, and killed many people. Like Australia, by the 1980s the consequences of removing natural fire from the landscape created dangerous fuel buildup. Prescribed fires and other forest-thinning projects in the United States are now frequently used to control the fuel buildup created from suppression for so many years.

In the Mediterranean also, historically fire was not trusted. People made every attempt to suppress and remove fire from the ecological cycle. Fire was not socially acceptable and was shunned as often as possible. Pyne (1997) writes that "Free-burning fire was the prerogative of free-ranging peoples, groups who wandered outside the fixed social order or who, by their mobility, threatened to destabilize that order and the landscape on which it depended" (p. 91). The social perception of fire during the late 1970s was such that, "In most Mediterranean countries, anyone considering the possibility that fire could be beneficial was in danger of being regarded as a pyromaniac and jeopardizing his professional career" (Naveh, 1990,

p. 2). This aversion to fire, natural or human caused, resulted in the Mediterranean region seeing more frequent fires and increasingly more devastation from those fires (Stocks and Trollope, 1992). Naveh (1990) wrote that it was not until 1989, that H. Biswell, an ecologist and forester, proved that fire could be used through prescribed burning for effective vegetation management.

In situations throughout the world, suppression of natural fire drastically altered the composition and structure of vegetation that relied on frequent fires. Many land managers learned through difficult lessons that fire is an essential part of the ecosystem. As more people relocate near or in forested land, anthropogenic causes (e.g., arson, campfires, etc.) result in more wildfires. The juxtaposition of human life, property, and wildlands creates a need for accurate mapping of fire activity and its impacts.

This chapter focuses on remote sensing technologies used to create accurate maps of burn scars and severity. Many band ratios, algorithms, and methods have been studied to meet these ends. This chapter discusses a number of commonly used methods, including KT (Kauth-Thomas), PCA (principal components analysis), neural networks and object-oriented classification, NDVI (normalized difference vegetation index), and NBR/dNBR (normalized burn ratio/differenced normalized burn ratio). Following the discussion of methods, the rationale for using the NBR/dNBR is made, and the results of an operational example are presented. The institutions and processes for making various stages of pre-, during, and postfire maps available to fire managers is also presented.

ACTIVE AND POSTFIRE MAPPING EFFORTS

The effects and degree of disturbance on ecosystems by fire can typically be assessed during two stages of fire activity: active and postfire. A variety of tools is available to measure fire effects. These tools are typically available at broad (national or continental) and local (fire incident) scales. Active fire detection technology is used by wildland fire managers as a tool to assist strategic firefighting and planning and as a land management tool. Mapping of burn condition after a fire is often used to quantify the impacts of the fire on a variety of environmental variables, such as soil condition, property damage, and water quality.

ACTIVE FIRE DETECTION

Active fire detection and mapping efforts play an important role in wildland fire suppression efforts. Fire detection technology allows managers and scientists to monitor the intensity, duration, and progression of active fires. This information is used for strategic and tactical planning and management of wildland fire suppression resources. Active fires are mapped at broad and local scales.

Broad Scale

The application of remote sensing systems to map actively burning lands has been the subject of many studies as there is a significant need to identify and characterize active fires. At a broad scale, a number of satellite remote sensing systems are currently used to map active fires. Satellite remote sensing systems that provide broad-scale imagery typically provide excellent temporal resolution but lack sufficiently fine spatial resolution. Two common satellite remote sensing systems used in broad-scale active fire detection are AVHRR (Advanced Very High Resolution Radiometer) and GOES (Geostationary Operational Environmental Satellites), both operated by the U.S. National Oceanic and Atmospheric Administration. Originally launched as weather monitoring instruments, AVHRR and GOES are utilized by scientists to map fires. The Hazard Mapping System, used for both AVHRR and GOES, is an integrated system that combines automated fire detection algorithms with human interpretation to identify fire activity and smoke with multiple sensors (McNamara et al., 2002). The automated fire detection algorithms used with these sensors are quite robust considering the capabilities of the individual sensors. For example, Li et al. (2000) compared the results of AVHRR 1-km fire detection algorithms compared to burn areas delineated from conventional aerial surveillance and satellite-based techniques. These results indicate that the AVHRR-based algorithms sufficiently detected active fires, at the expense of increasing the number of false detections, which is caused by AVHRR's low saturation threshold in its thermal bands, causing warm land features to be detected as fire. Feltz et al. (2003) reported on validation efforts for the wildfire automated biomass burning algorithm used with GOES imagery. The GOES sensor, with a spatial resolution of 4 km, consistently detects active large fire locations, especially in equatorial regions of the world. Due to higher satellite view angles, more errors occur when detecting active fires away from the equator (i.e., boreal regions).

Another satellite sensor used frequently to map active fires is MODIS (MODerate resolution Imaging Spectroradiometer) aboard NASA's (National Aeronautic and Space Administration) Terra and Aqua satellites, launched in 1999 and 2002, respectively. The MODIS system extends the fire detection and mapping capabilities of AVHRR and GOES and provides improved fire detection and mapping capabilities. In addition, MODIS was developed to create integrated measurements of the land, water, and atmosphere, including bands tailored to fire monitoring (Justice et al., 2002; Kaufman et al., 1998). The U.S. Department of Agriculture (USDA) Forest Service, in conjunction with NASA and the University of Maryland, uses MODIS data to create Web-based interactive maps showing active fire locations (http://active-firemaps.fs.fed.us and http://rapidfire.sci.gsfc.nasa.gov). This provides scientists, wildland fire managers, and the public a frequently updated, synoptic view of the fire situation at a broad scale. This information is used by wildland fire managers and the public and is useful in areas with large wilderness areas, such as Alaska and northern Canada.

Local Scale

Where more detail is needed, a different suite of remote sensing tools is available. One of the resources available to local wildland fire managers is the use of airborne thermal infrared fire-mapping systems. The process of extracting data from these systems is referred to as *infrared interpretation*. Wildland fire managers can task airplanes with high spatial resolution thermal infrared sensors to image the fire to record actively burning locations. These data can be rapidly transferred back to the wildland fire suppression teams by transmitting the digital images to receiving stations on the ground or by manual transfer methods. Airborne thermal infrared mapping systems create highly accurate maps of active fire fronts and potential dangers. This technology allows fire managers to focus energies quickly on the most sensitive areas. After fires are contained, the focus of mapping efforts shifts from active fire detection to burn scar mapping and capture of postfire effects.

POSTFIRE MAPPING

There is a significant need for accurate postfire data sets describing fire location, size, and severity. Resource managers frequently use maps of cumulative burns throughout their land base to update inventory and create management plans. Another postfire information need is burn severity maps. As fires burn in varying levels of severity throughout the burn scar, rehabilitation managers need to focus efforts on the most severely burned locations. For both needs, remote sensing technology allows scientists quick access to the required information.

Broad Scale

Satellite instruments available for broad-scale mapping have been used extensively for burn scar mapping. Burn scar data are utilized for several applications, including aerosol emission estimates, carbon cycle modeling, hazard assessment, resource management, and policy creation. In Spain, Vazquez et al. (2001) created burn scar maps showing fire sizes using images captured by the IRS (Indian Remote Sensing) satellite Wide Field Sensor and Linear Imaging Self-Scanner sensors. SPOT (Système pour l'Observation de la Terra) VGT (Vegetation) has also been utilized to map burn scars in Australia (Graetz et al., 2003) and Africa (Brivio et al., 2003). Fraser et al. (2000) and Fraser et al. (2003) used SPOT VGT multitemporal data to create a procedure for continental-scale mapping of boreal forests in Canada. Amiro and Chen (2003) not only used SPOT VGT imagery to map burn scars, but also used the imagery to determine historical burn scar age across Canadian boreal forests. MODIS is also regularly utilized to make broad-scale maps of burned areas (Martin et al., 2002; Justice et al., 2002; Roy et al., 2003), as well as AVHRR (Al-Rawi et al., 2001; Barbosa et al., 1998; Domenikiotis et al., 2003; Fernandez et al., 1997; Kasischke and French, 1995). In addition to coarse spatial resolution sensors, relatively fine spatial resolution Landsat data have been used to create general fire disturbance maps over large areas (Bowman et al., 2002; Koutsias and Karteris, 2000; Kramber, 1992; Sa et al., 2003; Zarriello et al., 1995).

Burn scar mapping using low-to-moderate spatial resolution imagery is useful to evaluate the location and extent of burned areas and to update resource inventories. For example, during the summer of 2004, wildfires burned more than 4.4 million Ha in Alaska and the Yukon Territory. Many of the fires were in rugged, remote, inaccessible areas. At a regional and national scale, fire and resource managers used moderate-to-coarse spatial resolution remotely sensed data to map the burn scars and update forest inventories. The imagery and methods used to create burn scar maps also assist in the monitoring of aerosols present in the air, the creation of carbon models, in making geologic hazard assessments, such as landslide potential, and in the creation of policy and management strategies.

Local Scale

When greater detail about a burn scar is desired, another suite of sensors is required to assess fire effects. Analysts use postfire imagery to estimate burn severity in terms of soil and vegetation. One method for creating these maps is the use of aerial photography. Aerial photography (digital or emulsion-coated film) provides excellent spatial resolution and results in detailed images of burned areas. These images are increasingly acquired and delivered in a digital format, which requires some image-processing skills, such as the ability to georeference and terrain correct the images (Bobbe et al., 2001). Because aerial photography usually covers a relatively small area, many photographs are acquired, requiring subsequent correction and mosaicking. The raster images can be added as a spatial layer into a geographical information system (GIS) to depict postfire condition accurately and to overlay with other GIS layers. Many cameras used for aerial photography can also acquire imagery using a color-infrared band combination. Instead of the typical red-green-blue acquisition, many cameras now include the NIR (near-infrared) band in place of the blue band. The NIR band is especially sensitive to vegetation health, making it possible for analysts to efficiently locate areas stressed by fire or other disturbances, such as insect infestation (Hall et al., Chapter 4, this volume). Typically, using aerial photography as part of a GIS for fire mapping is only feasible on smaller (<1000 Ha) fires due to the large number of photographs that would require correction and mosaicking. For instance, for a fire about 800 Ha, 35 photos of the appropriate scale would be required to cover the entire burned area and produce imagery with 1×1 m pixel size.

The use of satellite imagery to map postfire condition has become an increasingly common technology available to resource and rehabilitation managers. For example, sensors such as Landsat-5 TM (Thematic Mapper), Landsat-7 ETM+ (Enhanced Thematic Mapper Plus), ASTER (Advanced Spaceborne Thermal Emission and Reflection Radiometer), and SPOT-4 and -5 are widely used. These sensors provide multispectral imagery that includes the visible, NIR, shortwave infrared (SWIR), and thermal infrared bands at spatial resolutions appropriate for mapping of the landscape at this scale. Most sensors acquiring imagery at this scale provide a large footprint, from 60×60 km (SPOT and ASTER) to 185×185 km (Landsat). Unlike aerial photography, one image from a satellite sensor is often sufficient to cover an entire burned area. Plus, the satellite imagery is delivered in a digital format and may already be in a georeferenced form.

BURN PATTERNS AND SEVERITY

Remote sensing technologies can be utilized to delineate burn patterns and varying levels of severity. Burn severity can often be characterized by intensity and residence time of the fire. Intensity and residence time within a fire area are dictated by a host of environmental factors, such as weather conditions, slope, prefire vegetation condition and type, and ladder fuels. For example, some areas burn very hot for long periods of time, destroying virtually all vegetation in the area and causing extensive damage to the soil. Other areas may have only been visited by the fire for a short time, killing the undergrowth and then moving along.

Depending on the audience, *burn severity* means different things. For silviculturists, burn severity relates to timber or vegetation mortality. However, for soil scientists and hydrologists, burn severity relates to the postfire soil condition. Parsons and Orlemann (2002), soil scientists, defined burn severity as the degree of change caused by the fire and measured in terms of soil hydrologic functions. Davis and Holbeck (2001) also defined burn severity as a measure of how it relates specifically to effects of the fire on soil conditions and hydrologic function. While satellite imagery only measures first-order effects of a fire (such as vegetation mortality), soil scientists and hydrologists utilize correlations between satellite reflectance and the postfire soil properties. Augmented with field information, satellite reflectance data can be processed to create burn severity maps that show postfire soil condition.

Assessment of postfire condition is a function of resource management requirements. There are two kinds of assessments that are commonly performed to map postfire condition: emergency and extended. The purpose of each is dramatically different and reflects the needs of the stewards of the local area.

EMERGENCY ASSESSMENT

Because fire can dramatically alter a landscape, many considerations must be taken into account regarding stabilization and rehabilitation. In the United States, land management agencies deploy BAER (Burned Area Emergency Response) teams to "prescribe and implement emergency treatments to minimize threats to life or property or to stabilize and prevent unacceptable degradation to natural and cultural resources resulting from the effects of a fire" (USDA Forest Service, 2004, p. 17). This task is important due to the danger that fires and burned areas create for years to come. In areas of high burn severity, the land is susceptible to mud and debris slides during and after every rain event. BAER teams not only have to locate those areas of high burn severity, but also take into account the possible damage that could be created downstream by the mud or debris slides. Team members must consider such things as personal property, threatened and endangered species, archeological sites, water supplies, and threats to soil productivity. Emergency assessments are meant to be used to help direct emergency stabilization and rehabilitation. The timing is also significant: Emergency assessments are done as soon as possible after fire containment, typically between seven and ten days.

EXTENDED ASSESSMENT

Extended assessments highlight areas affected by the fire not discernible in the emergency assessment. These assessments are more directed at resource management. In some instances, vegetation may be severely stressed by a fire but not die for a few months. In other cases, severely stressed vegetation may recover and not die. These delayed vegetation reactions are best monitored by an extended assessment. These assessments are typically done during the first growing season following the fire event, making it possible to differentiate between healthy growing vegetation and dead vegetation. Extended assessments are useful for resource managers as they seek an accurate depiction of vegetation condition following a wildfire as well as an indicator of vegetation recovery.

COMMON BURN-MAPPING APPROACHES

With increases in fire size and severity throughout the United States and Canada, fire managers are increasingly reliant on satellite imagery to map and characterize postfire burn condition. During 2002, the USDA Forest Service's RSAC (Remote Sensing Applications Center) provided satellite imagery support to 73 incidents. During that summer, the average size of the fire supported by BAER teams on U.S. Forest Service lands was 15,000 Ha, including 5 larger than 40,500 Ha. Large fires are very difficult to map via aerial reconnaissance and ground observations. Increasingly, resource managers are utilizing remote sensing technology to map postfire burn condition. A number of methods and techniques have been researched and used in this effort, including KT tasseled cap, PCA, neural networks, and object-oriented classification, NDVI, and NBR/dNBR.

KAUTH-THOMAS TRANSFORM

One technique used for burn area mapping is the KT transformation, also known as the tasseled cap transformation (TCT). Kauth and Thomas (1976) created coefficients for the transformation based on Landsat Multi-Spectral Scanner (MSS) data to create a four-dimensional space from the original four MSS bands. The four new axes represent a soil brightness index, green vegetation index, yellow stuff index, and non-such index. Nearly all the burn area information can be found in the first two indices, brightness and greenness (Jensen, 1996). Building on this pioneering work by Kauth, Thomas, and others, new coefficients for Landsat TM and ETM+ sensors have since been created, often with indices called brightness, greenness, and wetness (Crist and Cicone, 1984).

Patterson and Yool (1998) used the KT in an attempt to map fire-induced vegetation mortality. Compared to another linear transformation, PCA, the KT performed reasonably well. Henry and Yool (2002) used the KT, among other transformations, to derive landscape metrics useful in mapping burn severity. Liebermann et al. (2004) used the KT on Landsat and IKONOS imagery to map fire effects in California on a Mediterranean climate, chaparral-dominated ecosystem.

PRINCIPAL COMPONENTS ANALYSIS

PCA is another transformation used to analyze remotely sensed data. PCA is described as a "dimensionality reduction technique that maps image data into a new and uncorrelated coordinate system" (Pereira et al., 1997, p. 156). Nearly all the multispectral image variance is shown in the first two axes, or components. Brewer et al. (2005) compared PCA with several image transformations for mapping burn areas. Using all 14 bands from the pre- and postfire Landsat images, PCA showed that Components 2 and 5 were the most correlated to burned areas. Pereira et al. (1997) also found high correlation to burns in Component 2. Single-scene PCA analysis has also been utilized successfully to map and characterize burn areas (Patterson and Yool, 1998).

NEURAL NETWORKS AND OBJECT-ORIENTED CLASSIFICATION

Specialized image analysis software packages, including *Definiens eCognition* (www.definiens-imaging.com) and Visual Learning System's *Feature Analyst* (www.featureanalyst.com) are used to map and characterize burn areas. These software packages use object-oriented classification techniques and machine learning technology, respectively, taking not only spectral characteristics into account, but also spatial patterns. *eCognition* uses knowledge-driven and data-driven information to create classifications. For training data, it can use many image-processing derivatives, including KT, PCA, NDVI, multispectral and panchromatic imagery, NBR, and other layers relating to topography, such as Digital Elevation Model, slope, and aspect. Based on the training data and the other layer inputs, *eCognition* creates a classification based on pattern recognition to produce objects as opposed to use of more common pixel-based models or classifications. Mitri et al. (2002) used *eCognition* to create models for burned area mapping on the Mediterranean island of Thasos. *eCognition* was used to map burn/nonburn and degree of burn. The results from this study showed burn/nonburn could be mapped with a strong agreement with a field-derived fire perimeter (98.85% overall classification accuracy).

An additional commercially available software package to perform object-based classifications on burned areas is Visual Learning System's *Feature Analyst* (FA). This software uses machine learning technology to create a classification. Users apply training data to "teach" the software what the user is interested in mapping; the software tries to find all instances of that object and then produces reports showing an extraction of all those similar features. The user then is able to approve or reject the results. The user can apply more training data or adjust the existing training data until a suitable result is obtained (Vanderzanden and Morrison, 2002). Brewer et al. (2005) worked with the developers of FA and tested the logic and algorithms now used in FA to map burn severity on fires in eastern Montana. Multitemporal analyses with FA provide excellent opportunities for consistency with local or regional vegetation maps.

NORMALIZED DIFFERENCE VEGETATION INDEX

The NDVI is an index that highlights chlorophyll absorption and NIR reflectance often indicative of vegetation content and vigor (Jensen, 1996). NDVI has been used as indicative vegetation biomass and has been commonly applied to map burn areas (Chuvieco et al., 2002; Pereira et al., 1997). Spectral reflectance curves for normal, healthy vegetation show a dramatic increase in reflectance of the NIR portion, while the adjacent red portion of the spectrum largely absorbs light (Jensen, 1996). This dramatic difference in spectral response is accentuated with NDVI, for which burned areas respond with an increase in red reflectance and a decrease in NIR reflectance. The algorithm is as follows:

$$NDVI = (NIR - Red)/(NIR + Red) \qquad (5.1)$$

The degree of change in reflectance is indicative of fire damage or effects. Henry and Yool (2002) used the NDVI and other vegetation indices to characterize fire-caused spatial patterns in Arizona. Sunar and Ozkan (2001) used the NDVI for a multitemporal analysis of a burn scar in Turkey. Using a Landsat TM scene as a prefire vegetation image and a postfire vegetation image from IRS-1C, NDVI was useful in characterizing the burned area.

NORMALIZED BURN RATIO

The NBR is another band ratio similar to the NDVI. In lieu of the red and NIR bands used in NDVI, NBR uses the reflectance from the NIR and midinfrared bands, also called the SWIR bands. NBR can be applied with multispectral imagery that has NIR and midinfrared bands; however, the ratio works best when using the midinfrared band centered at 2100 nm (Hudak et al., 2004). The algorithm is as follows:

$$NBR = (NIR - SWIR)/(NIR + SWIR) \qquad (5.2)$$

Healthy green vegetation reflects NIR energy. Conversely, NIR response decreases where the vegetation is sparse. Midinfrared energy is largely reflected by rock and bare soil, meaning that midinfrared band values will be high in bare, rocky areas with sparse vegetation and low in areas of healthy green vegetation. Imagery collected over a forest in a prefire condition will have high NIR band values and very low midinfrared band values. Imagery collected over a forest after a fire will have low NIR band values and high midinfrared band values.

López-Garcia and Caselles (1991) used a ratio of Landsat Bands 4 and 7 and found it to be a "good parameter for studying vegetation regeneration on burnt areas" (p. 36). Pereira et al. (1997) also discussed using the 4/7 ratio for burn mapping, mentioning that the midinfrared bands are less affected by atmospheric scattering, therefore reducing scattering at the surface. In addition, as many natural materials have a broader range of reflectance in the midinfrared bands than the visible bands, it is easier to differentiate between different cover types.

Key et al. (2002) studied the use of the 4/7 ratio (which they normalized and named the NBR) and undertook a comparison to other similar indices. Many researchers have used a change detection approach based on the NBR, called the dNBR in burn-mapping projects. The dNBR is simply an image differencing between a prefire NBR and a postfire NBR:

$$dNBR = NBR_{prefire} - NBR_{postfire} \qquad (5.3)$$

RATIONALE FOR THE USE OF THE NORMALIZED BURN RATIO

Hudak et al. (2004) compared field observations to dNBR and NDVI values for fires throughout the western United States that burned during 2003 and illustrated that dNBR values consistently described vegetation burn severity better than the NDVI. van Wagtendonk et al. (2003) performed a multitemporal analysis of dNBR values created from the Airborne Visible and Infrared Imaging Spectrometer and Landsat ETM+. Due to similar spectral capabilities, dNBR values from the Airborne Visible and Infrared Imaging Spectrometer were comparable to ETM+ results. Brown (2002) performed an analysis of the dNBR compared to traditional mapping techniques utilized by scientists involved in postfire rehabilitation, finding the dNBR to be a useful and accurate tool to map burn severity. Miller and Yool (2002) performed the dNBR on imagery of the Cerro Grande Fire in New Mexico and found this process more accurate than making a visual interpretation from high spatial resolution color infrared photography. Brewer et al. (2004) compared the dNBR to a number of burn techniques and found the dNBR to be the most practical tool for burn mapping and applicable across ecosystems.

Single-scene NBR and two-scene dNBR image classification methods generally produce overall accuracies in the range of 50–60% to delineate unburned, low, moderate, and high burn severity classes (Bobbe et al., 2004). Based on the needs of field teams in emergency assessments, this level of accuracy is acceptable. It is also important to note that the highest accuracies were observed in the high burn severity classes (67–74%). High burn severity classes also require the most focus and attention from rehabilitation managers to mitigate the fire effects on the multiple values at risk.

The USDA Forest Service and U.S. Geological Survey use the dNBR method to map postfire burn severity operationally as a support to BAER teams. Compared to other methods, the dNBR provides the best representation of postfire burn condition, and as a ratio it has proved applicable across ecosystems and regions. RSAC developed models to apply the dNBR method operationally on pre- and postfire imagery. Since 2001, RSAC has mapped nearly 2.5 million Ha of burned area for BAER teams as an aid to perform emergency assessment activities. BAER teams are dispatched to fire incidents to perform an emergency assessment highlighting areas of greatest soil burn severity. This assessment directs treatments to areas that would be most severely affected by the next rain event. As the timing of weather events can be variable, BAER teams are typically given between 7 and 10 days after fire containment to perform this assessment. Due to these limited time constraints of BAER teams, RSAC creates a dNBR image using the best-available satellite or airborne imagery. Typically, this is Landsat TM and ETM+. Due to the age of Landsat TM and recent issues with Landsat ETM+, other imagery sources are investigated on an ongoing basis. Alternate sensors that can be used operationally for this mapping process are ASTER and SPOT-4 or -5. With ASTER revisiting a location eight days after Landsat TM and SPOT having pointable functionality, suitable imagery can typically be acquired within the BAER team time constraints.

The dNBR requires a pre- and postfire image, resulting in issues for non-Landsat image types. For instance, having to build a prefire SPOT archive speculatively would be expensive. Also, SPOT-4 and -5 each contain a midinfrared band, but their spectral wavelength most resembles Landsat Band 5, not 7. Hudak et al. (2004) reported that the shorter midinfrared band in a ratio with the NIR band is inferior for burn mapping to the true NBR ratio using Landsat Bands 4/7. Because of these factors, SPOT imagery is used only when imagery from more spectrally appropriate sensors is not available. ASTER imagery contains midinfrared bands (Bands 4–9) spectrally similar to Landsat Band 7, making it a suitable substitute for Landsat TM or ETM+ as a postfire image. Thus, ASTER imagery is often used as a source of postfire imagery for BAER mapping.

BURN MAPPING CASE STUDY: THE WILLOW FIRE

Wildfires are a common occurrence in the southwestern United States. In 2004, in New Mexico and Arizona alone 12 major fires burned nearly 121,000 Ha. The largest of these fires, the Willow Fire, burned 49,080 Ha in central Arizona near the town of Payson, with suppression costs of over U.S. $9 million. Further compounding suppression efforts was that over 70% of the burned land was within a designated wilderness area. Designated wilderness areas on U.S. Forest Service lands are inaccessible to motorized travel. This lightning-caused wildfire started June 24, 2004, and burned across a wide range of elevations and vegetation types. Accessing and mapping this large burned wilderness area was difficult, which meant BAER personnel relied on remote sensing for support.

Daily fire growth was monitored on a broad scale by the MODIS Active Fire Mapping program. Maps were created based on MODIS image acquisitions several times daily. These maps showed fire progression, and image subsets were posted to the Internet for public viewing. Airborne thermal infrared mapping support was ordered to assist with detailed perimeter delineation and daily fire progression. Between the MODIS and airborne thermal infrared mapping support, the incident received daily updates on the fire condition based on remote sensing technology.

Near the end of the fire, BAER team leaders were tasked to perform an emergency assessment of the Willow Fire within ten days of fire containment. To make this assessment, they contacted RSAC for image support. RSAC created a dNBR image using a prefire Landsat-5 TM scene (acquired July 2, 2003; Figure 5.1) and the first usable postfire scene acquired over the burn area, a Landsat-5 TM scene, acquired on July 4, 2004 (Figure 5.2). The dNBR data set and satellite image subsets were posted to an FTP site for BAER team retrieval and used in initial mapping efforts. Unfortunately, the Willow Fire burned another 15,000 Ha after the postfire image was acquired; however, an adjacent, overlapping image was acquired on July 11. RSAC again created a dNBR image using the prefire scene and the new postfire scene from July 11, 2004. Both data sets proved helpful in the creation of the required burn severity maps for the incident (Figure 5.3).

CONCLUSIONS

Wildfires are a major disturbance on forested lands each year. While forest fire is a normal and healthy part of the ecosystem, the nearness of human life and property

FIGURE 5.1 Landsat-5 TM image showing the prefire (July 2, 2003) condition of the landscape. The town of Payson, Arizona, is at the top right of the scene.

to forested land creates new obstacles. Forest managers have to consider how the burned land will react to subsequent weather events, in terms both of runoff and erosion and of possible damage to human life and property. Because of this, mapping postfire characteristics is critical. Remote sensing provides an excellent resource for those involved in mapping fire effects. Remote sensing allows analysts to highlight quickly the most sensitive areas that require immediate rehabilitation in an effort to mitigate possible future damage.

Significant research has been performed on techniques to create maps of fire disturbance. The NBR is appropriate to meet a range of fire and burn severity mapping needs. Land management agencies in the United States currently use the NBR as an initial input in the creation of burn severity maps. The NBR is applicable across ecosystems, requires little a priori knowledge, and can be used operationally.

FIGURE 5.2 Landsat-5 TM image showing the postfire (July 11, 2004) condition of the landscape. Notice the clouds and related shadows in the image over the burn scar, a common operational problem when using satellite imagery.

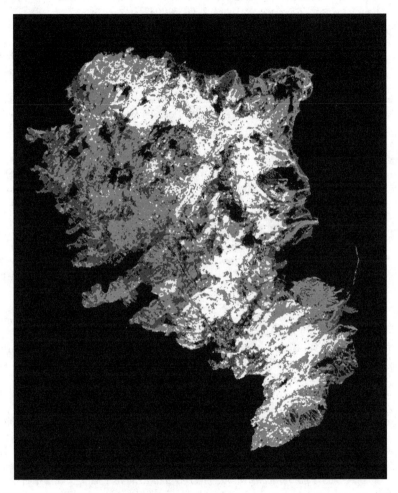

FIGURE 5.3 The dNBR classified into four classes. In this image, black represents unburned or very low underburn, while white represents high severity. BAER team members use this data set as a starting point in their emergency assessment. Approximately 10% of the land within the fire perimeter was unburned, 23% was low severity, 44% was moderate severity, and 23% was high severity.

REFERENCES

Al-Rawi, K.R., Casanova, J.L., and Calle, A. (2001). Burned area mapping system and fire detection system, based on neural networks and NOAA-AVHRR imagery. *International Journal of Remote Sensing, 22,* 2015–2032.

Amiro, B.D. and Chen, J.M. (2003). Forest fire scar aging using SPOT-Vegetation for Canadian ecoregions. *Canadian Journal of Forest Research, 33,* 1116–1125.

Barbosa, P.M., Pereira, J.M.C., and Grégoire, J.M. (1998). Compositing criteria for burned area assessment using multitemporal low resolution satellite data. *Remote Sensing of Environment, 65,* 38–49.

Bobbe, T., Finco, M.V., Maus, P., and Orlemann, A. (2001). *Remote Sensing Tools for Burned Area Emergency Response (BAER).* U.S. Department of Agriculture Forest Service, Remote Sensing Applications Center, Salt Lake City, UT. Report No. RSAC-43-TIP. 14p.

Bobbe, T., Finco, M.V., Quayle, B., Lannom, K., Sohlberg, R., and Parsons, A. (2004). *Field Measurements for the Training and Validation of Burn Severity Maps from Spaceborne, Remotely Sensed Imagery.* U.S. Department of Agriculture Forest Service, Remote Sensing Applications Center, Salt Lake City, UT. Final Project Report, Joint Fire Science Program-2001-2, RSAC-2001-RPT1. 15p.

Bowman, D.M.J.S, Zhang, Y., Walsh, A., and Williams, R.J. (2002). Experimental comparison of four remote sensing techniques to map tropical savanna fire-scars using Landsat-TM imagery. In *Proceedings from Fire and Savanna Landscapes in Northern Australia: Regional Lessons and Global Challenges* (pp. 341–348). CSIRO Publishing, Victoria, Australia.

Brewer, C.K., Winne, J.C., Redmond, R.L., Opitz, D.W., and Mangrich, M.V. (2005). Classifying and mapping wildfire severity: a comparison of methods. *Photogrammetric Engineering and Remote Sensing, 71,* 1311–1320.

Brivio, P.A., Maggi, M., Binaghi, E., and Gallo, I. (2003). Mapping burned surfaces in Sub-Saharan Africa based on multi-temporal neural classification. *International Journal of Remote Sensing, 24,* 4003–4018.

Brown, S. (2002). A Remote Sensing Based Assessment of Fire Severity. Unpublished master's thesis, University of Montana, Missoula, MT. 129p.

Chuvieco, E., Martin, M.P, and Palacios, A. (2002). Assessment of different spectral indices in the red–near-infrared spectral domain for burned land discrimination. *International Journal of Remote Sensing, 23,* 5103–5110.

Crist, E.P. and Cicone, R.C. (1984). Application of the tassled cap concept to simulated thematic mapper data. *Photogrammetric Engineering and Remote Sensing, 50,* 343–352.

Davis, M. and Holbeck, C. (2001). Nuts and bolts of BAER soil watershed assessments. In D. Harmon (Ed.), *Crossing Boundaries in Park Management. Proceedings of the 11th Conference on Research and Resource Management in Parks and on Public Lands* (pp.166–170). The George Wright Society, Hancock, MI.

Domenikiotis, C., Dalezios, N.R., Loukas, A., and Karteris, M. (2003). Agreement assessment of NOAA/AVHRR NDVI with Landsat TM NDVI for mapping burned forested areas. *International Journal of Remote Sensing, 23,* 4235–4246.

Feltz, J.M., Moreau, M., Prins, E.M., McClaid-Cook, K., and Brown, I.F. (2003). Recent validation studies of the GOES Wildfire Automated Burning Algorithm (WF_ABBA) in North and South America. *Proceedings of the Second International Wildland Fire Ecology and Fire Management Congress,* November 16–20, 2003. Orlando, FL. 6p.

Fernandez, A., Illera, P., and Casanova, J.L. (1997). Automatic mapping of surfaces affected by forest fires in Spain using AVHRR NDVI composite image data. *Remote Sensing of Environment, 60,* 153–162.

Fraser, R.H., Fernandes, R., and Latifovic, R. (2003). Multi-temporal mapping of burned forest over Canada using satellite-based change metrics. *Geocarto International, 18,* 37–47.

Fraser, R.H., Li, Z., and Landry, R. (2000). SPOT Vegetation for characterizing boreal forest fires. *International Journal of Remote Sensing, 21,* 3525–3532.

Graetz, R.D., Gregoire, J.M., Lovell, J.L., King, E.A., Campbell, S.K., and Tournier, A. (2003). A contextual approach to the mapping of burned areas in tropical Australian savannas using medium-resolution satellite data. *Canadian Journal of Remote Sensing, 29,* 499–509.

Henry, M.C. and Yool, S.R. (2002). Characterizing fire-related spatial patterns in the Arizona sky islands using Landsat TM data. *Photogrammetric Engineering and Remote Sensing, 68,* 1011–1019.

Hudak, A., Robichaud, P., Evans, J., Clark, J., Lannom, K., Morgan, P., and Stone, C. (2004). Field validation of Burned Area Reflectance Classification (BARC) products for post fire assessment. In J.D. Greer (Ed.), *Remote Sensing for Field Users: Proceedings of the Tenth Biennial Forest Service Remote Sensing Applications Conference* (unpaginated CD-ROM). American Society of Photogrammetry and Remote Sensing, Salt Lake City, UT.

Jensen, J.R. (1996). *Introductory Digital Image Processing: A Remote Sensing Perspective.* Prentice Hall, Upper Saddle River, NJ. 318p.

Justice, C.O., Giglio, L., Korontzi, S., Owens, J., Morisette, J.T., Roy, D., Desclortres, J., Alleaume, S., Petitcolin, F., and Kaufman, Y. (2002). The MODIS fire products. *Remote Sensing of Environment, 83,* 244–262.

Kasischke, E.S. and French, N.H.F. (1995). Locating and estimating the areal extent of wildfires in Alaskan boreal forests using multiple-season AVHRR NDVI composite data. *Remote Sensing of Environment, 51,* 263–275.

Kaufman, Y.J., Justice, C.O., Flynn, L., Kendall, J., Prins, E., Ward, D.E., Menzel, W.P., and Setzer, A.W. (1998). Potential global fire monitoring from EOS-MODIS. *Journal of Geophysical Research, 103,* 215–232.

Kauth, R.J. and Thomas, G.S. (1976). The tassled cap — a graphic description of the spectral-temporal development of agricultural crops as seen by Landsat. *Proceedings of the 1976 Symposium on Machine Processing of Remotely Sensed Data* (pp. 41–51). Laboratory for Applications of Remote Sensing, Purdue University, West Lafayette, IN.

Key, C.H., Zhu, Z., Ohlen, D., Howard, S., McKinley, R., and Benson, N. (2002). The normalized burn ratio and relationships to burn severity: ecology, remote sensing and implementation. In J.D. Greer (Ed.), *Rapid Delivery of Remote Sensing Products: Proceedings of the Ninth Biennial Forest Service Remote Sensing Applications Conference* (unpaginated CD-ROM). American Society of Photogrammetry and Remote Sensing, San Diego, CA.

Koutsias, N. and Karteris, M. (2000). Burned area mapping using logistic regression modeling of a single post-fire Landsat-5 Thematic Mapper image. *International Journal of Remote Sensing, 21,* 673–687.

Kramber, B. (1992). Fire scar mapping using Landsat multispectral scanner imagery. *Snake River Birds of Prey National Conservation Area Research and Monitoring Annual Report, 1992* (pp. 409–423). Department of the Interior, Bureau of Land Management, Boise District, Boise, ID.

Li, Z., Nadon, S., Cihlar, J., and Stocks, B. (2000). Satellite-based mapping of Canadian boreal forest fires; evaluation and comparison of algorithms. *International Journal of Remote Sensing, 21,* 3071–3082.

Liebermann, A., Stow, D., Rogan, J., and Franklin, J. (2004). Mapping burn severity in southern California Mediterranean type vegetation using Landsat and IKONOS data. In J.D. Greer (Ed.), *Remote Sensing for Field Users: Proceedings of the Tenth Biennial Forest Service Remote Sensing Applications Conference* (unpaginated CD-ROM). American Society of Photogrammetry and Remote Sensing, Salt Lake City, UT.

López-Garcia, M.J. and Caselles, V. (1991). Mapping burns and natural reforestation using thematic mapper data. *Geocarto International, 6,* 31–37.

Martin, M.P., Delgado, R.D., Chuvieco, E., and Ventura, G. (2002). Burned land mapping using NOAA-AVHRR and TERRA-MODIS. *Proceedings of IV International Conference of Forest Fire Research, 2002 Wildland Fire Safety Summit,* November 18–23, 2002. Luso, Coimbra, Portugal. 9p.

McNamara, D., Stephens, G., Ramsay, B., Prins, E., Csiszar, I., Elvidge, C., Hobson, R., and Schmidt, C. (2002). Fire detection and monitoring products at the National Oceanic and Atmospheric Administration. *Photogrammetric Engineering and Remote Sensing, 68,* 774–775.

Miller, J.D. and Yool, S.R. (2002). Mapping forest post-fire canopy consumption in several overstory types using multitemporal Landsat TM and ETM data. *Remote Sensing of Environment, 82,* 481–496.

Mitri, G.H., Gitas, I.Z., and Viegas, D.X. (2002). The development of an object-oriented classification model for operational burned area mapping on the Mediterranean island of Thasos using Landsat TM images. *Proceedings of IV International Conference of Forest Fire Research, 2002 Wildland Fire Safety Summit,* November 18–23, 2002. Luso, Coimbra, Portugal. 12p.

The Nature Conservancy. (2004). *Fire, Ecosystems and People: A Preliminary Assessment of Fire as a Global Conservation Issue.* Tallahassee, FL. Retrieved from http://nature.org/initatives/fire/files/fire_report_version1.pdf. 9p. March 15, 2005.

Naveh, Z. (1990). Fire in the Mediterranean — a landscape ecological perspective. In J.G. Goldammer and M.J. Jenkins (Eds.), *Fire in Ecosystem Dynamics: Mediterranean and Northern Perspectives* (pp. 1–20). SPB Academic Publishing, The Hague, The Netherlands.

Parsons, A. and Orlemann, A. (2002). Mapping post-wildfire burn severity using remote sensing and GIS. *Proceedings of the 22nd Annual ESRI International User Conference* (unpaginated CD-ROM), July 8–12, 2002. ESRI, San Diego, CA.

Patterson, M.W. and Yool, S.R. (1998). Mapping fire-induced vegetation mortality using Landsat Thematic Mapper data: a comparison of linear transformation techniques. *Remote Sensing of the Environment, 65,* 132–142.

Pereira, J.M.C., Chuvieco, E., Beaudoin, A., and Desbois, N. (1997). Remote sensing of burned areas. In E. Chuvieco (Ed.), *A Review of Remote Sensing Methods for the Study of Large Wildland Fires* (pp. 127–183). University of Alcala, Alcala de Henares, Spain.

Pyne, S.J. (1991). *Burning Bush: A Fire History of Australia.* Henry Holt and Company, New York. 552p.

Pyne, S.J. (1997). *Vestal Fire: An Environmental History, Told Through Fire, of Europe and Europe's Encounter with the World.* University of Washington Press, London. 408p.

Roy, D.P., Lewis, P.E., and Justice, C.O. (2003). Burned area mapping using multi-temporal moderate spatial resolution data — a bi-directional reflectance model-based expectation approach. *Remote Sensing of Environment, 83,* 263–286.

Sa, A.C.L., Pereira, J.M.C., Vasconcelos, M.J.P., Silva, J.M.N., Ribeiro, N., Awassee, A. (2003). Assessing the feasibility of sub-pixel burned area mapping in miombo woodlands of northern Mozambique using MODIS imagery. *International Journal of Remote Sensing, 24,* 1783–1796.

Stocks, B.J. and Trollope, W.S.W. (1992). Fire management: principles and options in the forested and savanna regions of the world. In P.J. Crutzen and J.G. Goldammer (Eds.), *Fire in the Environment: The Ecological, Atmospheric, and Climatic Importance of Vegetation Fires* (pp. 315–326). John Wiley and Sons, New York.

Sunar, F. and Ozkan, C. (2001). Forest fire analysis with remote sensing data. *International Journal of Remote Sensing, 22,* 2265–2277.

United States Department of Agriculture Forest Service. (2004). *Forest Service Manual 2500 — Watershed and Air Management.* U.S. Department of Agriculture, Forest Service. 44p. Washington, D.C.

Vanderzanden, D. and Morrison, M. (2002). *High-Resolution Image Classification: A Forest Service Test of Visual Learning System's Feature Analyst.* Department of Agriculture Forest Service, Remote Sensing Applications Center, Salt Lake City, UT. Report No. RSAC-3004-RPT1. 17p.

van Wagtendonk, J., Root, R.R., and Key, C.H. (2003). Comparison of AVIRIS and Landsat ETM+ detection capabilities for burn severity. *Remote Sensing of the Environment, 92,* 397–408.

Vazquez, A., Cuevas, J.M., and Gonzalez-Alonso, F. (2001). Comparison of the use of WiFS and LISS images to estimate the area burned in a large forest fire. *International Journal of Remote Sensing, 22,* 901–907.

Zarriello, T.J., Knick, S.T., and Rotenberry, J.T. (1995). Producing a burn/disturbance map for the Snake River Birds of Prey National Conservation Area. In *Snake River Birds of Prey National Conservation Area Research and Monitoring Annual Report, 1994* (pp. 333–343). U.S. Department of the Interior, Bureau of Land Management, Boise District, ID.

6 Integrating GIS and Remotely Sensed Data for Mapping Forest Disturbance and Change

John Rogan and Jennifer Miller

CONTENTS

INTRODUCTION

Scientists and policy makers from various institutions and agencies are currently devoting substantial time and resources to study the implications of environmental change in forests and woodlands, the most widely distributed ecosystem on the earth (McIver and Wheaton, 2005; Wulder, 1998). In the context of environmental remote sensing, forest change, manifested as forest attribute modification or conversion, can occur at every temporal and spatial scale, and changes at local scales can have cumulative impacts at broader scales (Loveland et al., 2002). Natural resource managers and environmental modelers thus require reliable information about the ecological impacts associated with natural and anthropogenic disturbances to forests (Bricker and Ruggiero, 1998; Mladenoff, 2005).

Current understanding of the extent and rate of forest change is inadequate because (a) long-term large-area monitoring, suited to mapping conversions and transitions, is in its operational infancy (S. E. Franklin and Wulder, 2002); and (b) modifications to forest condition/abundance are difficult to detect with reliable precision (Gong and Xu, 2003). As such, researchers and policy makers "lack ... quantitative, spatially-explicit and statistically representative data on land-cover change" (Lambin, 1999, p. 191). To redress this deficiency, the GIS science community has begun to explore new ways to detect, characterize, and monitor forest change through the integration of remote sensing and GIS (geographical information system) data and technologies (Kasischke et al., 2004).

The integration of remotely sensed and GIS data* takes four forms: (a) GISs can be used to store multiple data types; (b) GIS analysis and processing methods can be used for raster data manipulation and analysis (e.g., buffer/distance operations); (c) remotely sensed data can be manipulated to derive GIS data; and (d) GIS data can be used to guide image analysis to extract more complete and accurate information from spectral data. This chapter focuses on the fourth topic, with acknowledgment of the gains made by the academic GIS science community in the areas of spatial analysis, landscape conceptualization, and map validation (National Center for Geographic Information and Analysis [NCGIA], 2005).

GIS data, such as topographic variables, were first integrated in remote sensing-based vegetation mapping studies in the late 1970s because available satellite data (e.g., Landsat Multi-Spectral Scanner) did not provide sufficient floristic detail for effective resource management (see Franklin, 1995, and references therein). For many applications, this problem is still current. The rationale for incorporating topographic variables in single-date forest mapping is based on their correlation with forest species or lifeform composition (Guisan and Zimmerman, 2000). For example, the Utah Gap

---

* *GIS data* in this chapter refers to all nonspectral digital entities included and used in forest mapping/monitoring applications. GIS data have also been described as "collateral" and "ancillary" as compared to "primary" remotely sensed data (Jensen 2005).

Analysis Program (GAP) land cover mapping program uses topographic data (elevation, slope, aspect, and position index) and soil data (carbon content, available water, and quality) to produce fine-scale vegetation maps (C. Huang et al., 2003). In spite of the substantial improvement in remote sensing technology and data quality (spatial and spectral resolution) since the 1970s, however, the need for contextual GIS data has actually increased because remote sensing scientists are posing more complex questions than ever before (e.g., land change science; Turner et al. 1999; Turner, 2002). Including GIS data with remotely sensed data for the discrimination of land cover classes typically results in higher overall map accuracies (e.g., increases of 5–10% overall) over those produced using spectral-radiometric data alone (Frank, 1988; Senoo et al., 1990; Strahler et al., 1978; Talbot and Markon, 1986; Trietz and Howarth, 2000). For a review of recent advances in land cover mapping, see the work of S. E. Franklin and Wulder (2002) and J. Franklin et al. (2003).

Forest change mapping and monitoring is feasible when changes in the forest attributes of interest result in *detectable* changes in image radiance, emittance, or microwave backscatter values (Coppin et al., 2004). Forest disturbances vary by type, duration, and intensity (Gong and Xu, 2003). Disturbances such as wildfire, insect infestation, disease, timber harvest, ice storms, flooding, and strong winds usually result in highly variable (spectrally and spatially) damage at scales ranging from leaves to landscapes (Attiwill, 1994). Accurate remote sensing assessment of disturbance impact, severity, and rate of recovery (succession) can therefore be difficult and even impossible in substantially heterogeneous landscapes in relation to the sensor's spatial, spectroradiometric, and temporal characteristics (Rogan and Chen, 2004). When *known* or *perceived* changes to forest attributes occur but cannot be detected, located, or characterized to an acceptable confidence level, GIS data can thus play an important role in facilitating more robust change mapping (Rogan et al., 2003).

Change detection analysis employing both GIS coverages and remotely sensed images obtained prior to and following a disturbance has been used to assess specific types of forest and woodland damage, including vegetation cover responses to drought (Jacobberger-Jellison, 1994; Peters et al., 1993); insect outbreaks (S. E. Franklin et al., 2003; Nelson, 1983); windthrow (Cablk et al., 1994; Johnson, 1994); ice storm impacts (Olthof et al., 2004); and timber harvest (Nepstad et al., 1999; Sader et al., 2003). In many of these studies, the integration of spectral and GIS data was shown to improve substantially impact/damage assessment and map accuracy.

The objectives of this chapter are to describe how GIS data and technology can be utilized as a tool to characterize forest disturbance and change, how GIS data can be used to complement remotely sensed data, and how they can be used together to map and model forest conversions and modifications.

INTEGRATION OF GIS AND REMOTELY SENSED DATA

Complete integration of remotely sensed and GIS data is a long-standing problem that has drawn the attention of the International Society of Photogrammetry and Remote Sensing (Commission IV) and the (U.S.) NCGIA (Initiative 12). The integration of GIS data with remotely sensed imagery has witnessed increased interest for the following reasons:

1. Increased data availability, quality, and decreased data costs across large study extents (Davis et al., 1991; Emch et al., 2005; Treitz and Rogan, 2004)
2. Development of large-area forest mapping/monitoring projects using a wide variety of spectral data captured by different platforms, featuring disparate spatial and spectroradiometric characteristic capabilities (e.g., MSS vs. Advanced Spaceborne Thermal Emission and Reflection Radiometer) (Franklin and Wulder, 2002)
3. Demand for more precise estimates of disturbance impacts with Landsat-like data (i.e., spatial and thematic resolutions) (Seto et al., 2002; Varjo, 1997; E. H. Wilson and Sader, 2002)
4. Growing need for automated mapping and map updating in complex landscapes using expert systems/knowledge-based classification (X. Huang and Jensen, 1997; Lees and Ritman, 1991; Raclot et al., 2005)
5. Demonstrated potential of data integration/fusion for predictive forest change mapping (Baker, 1989; Mladenoff, 2005; Rogan et al., 2003)

Gahegan and Flack (1999) stated that the relationship between remote sensing and GIS has traditionally been that of supplier (remote sensing) and consumer (GIS). Typical remote sensing-derived products used in GIS analyses include baseline forest cover and lifeform maps (S. E. Franklin, 2001) and forest cover change maps used for map updating (Levien et al., 1999; Zhan et al., 2002); these are available at spatial resolutions typically ranging from 10 m to 1 km. The current spatial and spectral capabilities and limits of baseline mapping for generic change detection are discussed in detail in the work of J. Franklin et al. (2003) and Rogan and Chen (2004). In addition, digital elevation models (DEMs) can be generated using a variety of sensors and established methodologies (S. E. Franklin, 2001). The primary methods for DEM production are stereogrammetric techniques using air photos (photogrammetry), optical spaceborne imagery (SPOT [Systeme Pour l'Observation de la Terre] and Advanced Spaceborne Thermal Emission and Reflection Radiometer), and radar data (interferometry). Airborne light detection and ranging (LIDAR) data have been applied to terrain mapping. While LIDAR-derived DEMs have fine spatial resolution and high horizontal and vertical accuracy, currently they do not offer widespread coverage (Jensen, 2005; Lim et al., 2003), with some regional exceptions such as the Puget Sound Lidar Consortium (http://rocky2.ess.washington.edu/data/raster/lidar/index.htm). In addition, the extraction of linear features such as roads, trails, and streams using high spatial resolution optical data has reached a high level of sophistication and automation (Song and Civco, 2004).

Remote sensing analysts have become avid consumers of GIS data as a means to add value to remotely sensed data and analysis (S. E. Franklin, 2001). While there are many superficial similarities between GIS and remotely sensed data, a few conceptual differences make the complete integration of GIS and remote sensing challenging. Dobson (1993) noted two chief problems related to remote sensing and

GIS integration, such as incompatible data types (e.g., DEMs and census data) and the lack of an integrated approach to spatial data handling. Lees (1996) noted that the separate operational data spaces of spectral and spatial (GIS) variables must be acknowledged to conduct meaningful analysis. GIS data space is defined by the values of the direct/indirect variable of interest (e.g., temperature, elevation), while spectral data space consists of a discrete slice of the electromagnetic spectrum, which the remote sensing community needs to address further (Lees, 1996). Gahegan and Ehlers (2000) discussed the transformation process from a remotely sensed image to classified theme to subsequent GIS object and the error propagated at each step. Gahegan and Flack (1999) added that benefits to more seamless integration include the potential for more specific, and therefore more meaningful, data products and the ability to use GIS products to provide typicality information (e.g., ecological structure and function) as well as ancillary data to add more information to remotely sensed products (see also Aspinall, 2002).

GIS data are integrated in forest cover mapping and monitoring in three primary ways (Hutchinson, 1982):

1. Preclassification stratification — partitioning the study area based on elevation gradients or watershed boundaries to minimize the number of spectral classes or separate classes that are spectrally similar but geographically distinct (Cibula and Nyquist, 1987; J. Franklin et al., 1986; Vogelmann et al., 1998). This method is particularly relevant in forest disturbance contexts to mask either irrelevant or confounding scene features (Coppin et al., 2004)
2. Postclassification sorting — partitioning mapped categories based on soil type or slope to disaggregate or refine class membership* (Loveland et al., 2002; Satterwhite et al., 1984; Shasby and Carneggie, 1986). This method is in wide application in expert knowledge base approaches (X. Huang and Jensen, 1997)
3. Direct inclusion — combining ancillary variables with spectral data in a classification† (Ricchetti, 2000; Rogan et al., 2003; Wulder et al., 2004). This method is in increasing use with machine learning classification algorithms

The first two methods are analytical and assume that the analyst has "expert" knowledge of the study area and can therefore use environmental relationships in the ancillary data (e.g., slope) to stratify the remotely sensed data so they will be manipulated differently (e.g., one slope interval vs. another) via preclassification stratification. Postclassification stratification uses expert knowledge to aggregate map classes based on environmental relationships (e.g., vegetation classes can be stratified by elevation zones). Typically, the use of either continuous or discrete

---

* This may be applied to refine categories generated using supervised classification or to label unsupervised classes.

† Strahler (1980) also suggested the use of ancillary variables to calculate prior probabilities for the maximum likelihood classifier to improve map accuracy, and this has been implemented in large-area mapping using decision trees (McIver and Friedl 2002).

ancillary data is dependent on the classification technique used (Brown et al., 1993), with similar variables represented as discrete (i.e., separated categories at critical thresholds) or continuous (i.e., distance-based or interpolated coverage maps and surfaces) in accordance with the input requirements. The first two methods have therefore been used previously when parametric classification algorithms are employed because they are unable to handle categorical inputs directly (Strahler, 1980). Direct inclusion takes an empirical approach to mapping where the ancillary variables are included in the classification process with remotely sensed data potentially to provide additional information for improved class discrimination (Rogan et al., 2003). Wulder et al. (2004) stressed the need for data rescaling when DEM data are included with remotely sensed data involving parametric classification algorithms. Both continuous and discrete data are handled readily by nonparametric machine learning algorithms (MLAs), however (Rogan et al., 2003; Saveliex and Dobrinin, 2002).

GIS DATA AS ENVIRONMENTAL VARIABLES

The selection of input data for forest disturbance mapping and monitoring can have a significant impact on the final map product (Gong and Xu, 2003). Even when using relatively simple processing algorithms such as a minimum distance classifier, GIS data can facilitate detection and discrimination of target features, which could prove more beneficial than scarce or poor-quality input data processed with a complex algorithm. Biological, physical, and socioeconomic properties of the environment strongly influence land surface processes and human behavior and subsequently vegetation composition, abundance, and condition (Steyaert, 1996; Warner et al., 1994). This makes the selection and characterization of these variables increasingly important (Guisan and Zimmerman, 2002). Further, Skidmore (1989) noted that the relative importance of different types of GIS data can vary by spatial scale. For example, topographic data can improve land cover map accuracy at local to regional scales, whereas climate data become more important at regional to global scales.

GIS data that are potentially important in mapping, monitoring, and modeling forest change are described in Table 6.1. Variables that describe topography have been used in most environmental modeling applications (J. Franklin, 1995; Guisan and Zimmerman, 2000) as they are correlated with vegetation distribution at a finer spatial scale than climate variables (J. P. Wilson and Gallant, 1998; see Florinsky, 1998, for review of relationships between topographic variables and landscape characteristics). Simple topographic variables such as elevation, slope, and aspect most often represent indirect gradients (Austin and Smith, 1989) with respect to forest species distribution. Slope, however, can be considered a direct variable in the context of disturbance such as fire (i.e., slope steepness is directly related to flame length and burn intensity) (Rogan and Yool, 2001).

Figure 6.1 presents a conceptual diagram of a forest disturbance mapping/modeling scenario. The impact of abiotic and biotic disturbances on a forest stand is mostly determined by the interaction of the intensity of dynamic disturbances (e.g., wind speed) and their severity, or immediate impact, as mitigated/enhanced by static factors (i.e., topography), and the intrinsic properties of the forest stand (i.e., com-

TABLE 6.1
Potential GIS Data in Forest Change Studies

Variable type/source	Examples	Typical resolution	Type	Source	Association
Topography: http://seamless.usgs.gov/website/seamless/	Simple: elevation, slope, aspect	1:24,000 (~30 m)	Continuous, ordinal	DEM	Climate, soil characteristics, disturbance-behavior
	Complex: topographic moisture index, incoming solar radiation, landscape position		Continuous, ordinal, categorical	Formulae using DEMs	Soil texture, available moisture, temperature
Climate: http://www.wcc.nrcs.usda.gov/climate/prism.html	Minimum January temperature, maximum July temperature, summer precipitation	~4 km	Continuous	Interpolated weather station data	Available moisture
Disturbance: http://glcf.umiacs.umd.edu/data/modis/vcc/index.shtml	Wildfire burn scar perimeters Insect pest infestation perimeters Ice storm damage intensity perimeters Timber harvest removal/type perimeters	~1 Ha (MMU)	Categorical, ordinal	Compiled, field work, aerial sketch/survey	Disturbance process (type and intensity/severity)
Forest inventory: http://fia.fs.fed.us/	Stand age, species, dbh, density, height, crown bulk density	Variable	Categorical, ordinal	Compiled, field work	Species/stand abundance, condition, composition
Socioeconomic: http://www.census.gov/geo/www/index.html		1:100,000	Categorical	U.S. Census	Anthropogenic influence/policy

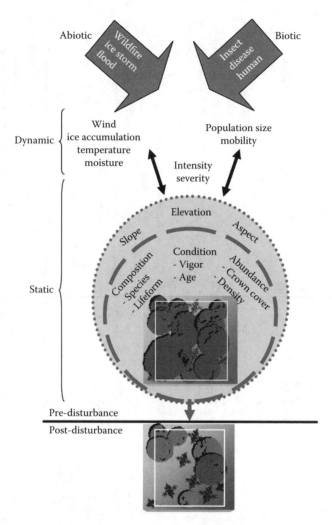

FIGURE 6.1 Forest disturbance mapping/modeling paradigm.

position, condition and, abundance). Because frequent and accurate data representing dynamic drivers of disturbance are rare and costly to collect, readily available static variables (e.g., elevation and slope) are more often used in conjunction with remotely sensed data to map both disturbance and disturbance-risk (Medler and Yool, 1997; Gemmell et al., 2002). For example, Khorram et al. (1990) found the indicator of conifer forest decline, as defined by the percentage of defoliation, to be a function of Landsat-5 Thematic Mapper (TM) (near infrared), elevation, and aspect.

    Complex topographic variables (e.g., topographic moisture index, incoming solar radiation) can have more direct influence on forest distribution and usually describe a combination of factors such as soil texture and water availability associated with the microclimate of a location (see Moore et al., 1991). The U.S. Geological Survey provides a standard 1:24,000 (7.5-min) digital DEM data set for the conterminous

United States (10-m DEM coverages are also available for some areas; see http://edc.usgs.gov/products/elevation/ned.html). In 2000, the Shuttle Radar Topography Mission obtained elevation data on a near-global scale to generate the most complete high-resolution digital topographic database of Earth. The University of Maryland Global Land Cover Facility editions of Shuttle Radar Topography Mission data are available in three general formats: 1 arc-sec (30 m) of the United States; 3 arc-sec (90 m) of the world; and 30 arc-sec (1 km) of the world (http://www.land-cover.org) (see Table 6.1).

Climate has a direct influence on forest distribution, typically through extremes in temperature and precipitation amounts. Some of the earliest fine spatial resolution (4-km) climate maps were produced through collaboration between Natural Resources Conservation Service National Water and Climate Center and Spatial Climate Analysis Service at Oregon State University. Based on a model named PRISM (parameter-elevation regressions on independent slopes model), factors such as rain shadows, temperature inversions, and coastal effects were incorporated in the climate-mapping process (see http://www.ocs.orst.edu/prism/). Liu et al. (2003) included elevation and temperature variables to map the entire land cover of China.

Existing land cover and land use data can be crucial for land cover stratification and vegetation sampling and analysis. The historical legacy of a particular land use on vegetation distribution has been examined (Foster et al., 1998). For example, Pan et al. (2001) reported that physical attributes explain only a small portion of the abundance of conifer species located on past abandoned land compared to land use factors. The U.S. Geological Survey provides a land use and land cover data set with 21 possible cover type categories, based primarily on manual interpretation of 1970s and 1980s aerial photography (see http://edc.usgs.gov/products/land-cover/lulc.html). Existing vegetation maps, however, are currently considered too coarse for detailed analyses (Coulter et al., 2000, p. 1329). Soil type provides information on texture, moisture and nutrient availability, and pH and has been used to prestratify a boreal forest based on mineral soil type to reduce the influence of soil background variation of timber harvest mapping (Heikkonen, 2004). Soil data, both spatial and tabular, are available from the (U.S.) Natural Resources Conservation Service.

Contextual socioeconomic data such as roads, distance to roads, and human population (census) have been directly linked to forest change, usually in the form of proximity to disturbance "potential" (Eastman et al., 2005). Chen (2002) discussed the limitations involved in selecting an appropriate scale to which census data should be disaggregated to be compatible with raster-based imagery (and GIS data). Mertens et al. (2001) combined ecological, economic, and remotely sensed data to predict the impact of logging activities on forest cover in east Cameroon. Results showed that the occurrence of logging-induced forest cover modifications increased with the value of forest rent.

Disturbance-related GIS data include burn scar perimeters, timber harvest polygons, and flood maps and have been used in a variety of ways, including:

1. Masking previous/other disturbances not of interest to the mapping exercise
2. Calibrating and validating classification algorithms

3. Testing the effectiveness of spectral change-thresholding procedures
4. Validating forest disturbance products derived from remotely sensed data

While extremely useful for change mapping studies, supporting GIS data are rarely collected in a repeated and consistent manner or the same spatial resolution. For example, in California, burn scar perimeters are available at 4-Ha minimum mapping unit (MMU) on U.S. Forest Service lands, while state forest lands provide 121-Ha MMU data. However, 500-m 16-day MODerate Resolution Imaging Spectroradiometer (MODIS) burn scar products have recently become available for large-area fire monitoring (see http://edcdaac.usgs.gov/modis/dataproducts.asp). Daily maps of active thermal hot spots are available from the MODIS 8-day 1-km active fire summary product data, indicating the spatial location of active fires (Roy et al., 1999).

Errors in GIS Data

No digital data set is error free. An essential condition for successful integration of GIS and remotely sensed data is an understanding of the error contained in the data and propagated in subsequent analyses (Hinton, 1999). The topic of uncertainty (i.e., a quantitative statement about the probability of spatial data error) is a central theme in GIS science literature. Accuracy assessment of land cover from remotely sensed data is a mature topic (Foody, 2002). In contrast, accuracy assessment of the results of change detection applications have received a relatively modest amount of attention in the remote sensing (change detection) literature (Woodcock, 2002). Categorical variables such as soil type or land use are prone to errors, such as positional, topological, and attribute inaccuracies. Guisan and Zimmerman (2000) noted the importance of high accuracy in categorical variables because they often act as "filters" for primary prediction when combined with continuous variables such as elevation. While categorical data may be perceived to be more "accurate" than remotely sensed representations when presented in vector format (e.g., crisp boundaries), they can often have a coarser minimum mapping unit than the image data (Coulter et al., 2000).

Continuous variables, specifically topographic variables, have special importance in forest change mapping because they are commonly used in the derivation of additional variables (see Hunter and Goodchild, 1997). The accuracy of topographic variables depends primarily on the accuracy of the DEM from which they were derived (Florinsky, 1998). Various studies have investigated the effect of error in DEMs on data derived from them (Bolstad and Stowe, 1994; Hunter and Goodchild, 1997; Lees, 1996). For example, slope computed from a DEM not only is affected by the algorithm used to derive it, but also is affected by the precision of the elevation values in the DEM (Perlitsch, 1995).

It has generally been accepted that, as the steps involved in derivation of a topographic variable increase, so does its susceptibility to error (Guisan and Zimmerman, 2000). Van Niel et al. (2004) noted that this is not always the case, however. In a study that simulated the propagation of error in topographic variables, they found that in some cases more complex variables such as net solar radiation were less affected by error than relatively simple variables such as slope and aspect (Van

Niel et al., 2004). Holmes et al. (2000) similarly found that topographic variables derived by compounding values from a large number of DEM grid cells were affected by errors most dramatically, and that while global error estimates may be low, their local error measurements could be quite high.

In summary, the key challenges associated with a more complete understanding of error in GIS variables involves a lack of procedures or protocol for quality control of integrated data (geometric accuracy and thematic detail) and issues related to different levels of data abstraction and representation (resolution and scale). Finally, a common, yet often unreported, issue is that dynamic GIS data representations such as land cover/use are out of date as soon as they are produced and may cause map errors when used for image masking, class sorting, or predictive modeling (Steyaert, 1996).

CONTRIBUTION OF GIS DATA TO FOREST CHANGE MAPPING

The typical forest change detection and mapping process consists of the following steps: (a) acquisition and coregistration of multidate imagery; (b) radiometric processing; (c) image transformation and change mapping; and (d) validation and change analysis (Coops et al., Chapter 2, this volume). GIS data are important for all four steps (Table 6.2).

DATA ACQUISITION AND COREGISTRATION

Data Acquisition

From the outset of a change-mapping project, GIS data can be used to delineate specific "mapping zones" (Homer and Gallant, 2001), such as geographic areas (political boundaries), biomes (ecoregions), topography (watersheds), and land cover/use/ownership. The use of mapping zones can serve to maximize spectral uniformity, provide boundary delineation, and partition the workload into "logical, feasible units" (S. E. Franklin and Wulder, 2002, p. 16). Ramsey et al. (1995) concluded that ecoregions could be characterized based on phenological variation of vegetation cover using normalized difference vegetation index distribution maps as surrogates for net primary productivity. Bergen et al. (2005) used major land resource areas defined by biophysical and socioeconomic constraints.

In many forestry applications, the "stand" is used as the minimum unit of analysis rather than the pixel (J. Franklin et al., 2003) because medium-resolution Landsat TM pixels often possess higher spatial resolution than the vegetation attributes under investigation. Image segmentation has thus been applied to image data to delineate forest stands (J. Franklin et al., 2000). At fine spatial scales, which may have specific geographic features for detailed study, rivers and roads are often used for landscape delineation. For example, Congalton et al. (2002) used stream buffers to aid identification and monitoring in a riparian forest. GIS data can also be used to delimit target features or "damage zones" on thematic scales as broad as "the damaged area"

TABLE 6.2
Contribution of GIS Data to Forest Change Mapping

Task	Contribution of GIS	GIS data	
		Continuous	Categorical
Data set selection	Biome delineation	Topography	Land/vegetation cover, land use, land ownership, political boundaries
	Study area boundary		Disturbance polygons
	Delimiting targeting features	n/a	Roads, streams
Image preprocessing	Geometric correction	DEM, GPS coordinates	n/a
	Orthorectification	DEM	n/a
	Terrain correction	DEM	
Image enhancement	Choice of transformation	DEM, slope	Vegetation, soils
Detection and mapping	Map legend(s)	n/a	Land cover/vegetation
	Stratification (masking); segmentation	DEM, slope, aspect	Soils, land/vegetation cover
	Change thresholding	n/a	Disturbance perimeters
	Calibration	Topography	Vegetation, soils, distance to roads[a]
	Choice of classification model (i.e., parametric vs. nonparametric)	All	All
	Validation	GPS coordinates	Vegetation, disturbance perimeters[a]
Change analysis	Cross tabulation/area summaries	Topographic variables	Land cover/use/ownership
Change modeling	Explanatory variables	All	All

[a] Extant.

n/a = not applicable.

to scales as fine as individual patches (e.g., wildfire burn scar and timber harvest plan perimeters) (e.g., Rogan and Yool, 2001; E. H. Wilson and Sader, 2002).

Geometric Correction

Image data acquired by satellite and airborne sensors are affected by systematic sensor, platform-induced, and terrain distortions that are introduced when image geometry is not perpendicular. Accurate per-pixel registration of multitemporal remotely sensed data is essential for forest change mapping because the potential exists for registration errors to be interpreted as forest cover change, which can lead to overestimation of actual change (Stow, 1999). Distortions in image data can be corrected by developing a model to tie per-pixel image features to GIS-based ground features (e.g., roads, streams, ridgelines [DEM], topographic maps). Further, in mountainous areas, terrain displacement can be hundreds of meters. For example, the 4-m multispectral image product from the IKONOS-2 will have nearly 600 m of terrain displacement if the sensor acquires data over an area with a kilometer of vertical relief where the sensor has an elevation angle of 60° (30° from nadir). To remove the terrain distortions accurately, DEMs are used to perform image orthorectification on optical and microwave data. Unfortunately, one of the shortcomings of current DEMs is that spatial resolution is often too coarse for orthocorrecting fine-resolution remotely sensed data such as QuickBird (2.44-m spatial resolution-multispectral) (Jensen, 2005).

Radiometric Processing (Terrain Correction)

DEMs and vegetation maps are commonly used in topographic normalization (terrain correction) of optical and microwave data. For optical data, terrain correction procedures are typically based on a model that adjusts the radiance values measured by a sensor using data depicting the local terrain (Smith et al., 1980). The Minnaert model is used for topographic normalization, so called because reduction of topographic effects in each image pixel is based on the generation of a normalized radiance value (i.e., the radiance that the pixel would have if the terrain within the scene was flat). Because the Minnaert approach does not assume that surface cover is a perfect diffuse reflector, it requires the calculation of a photometric (Minnaert) constant K that is specific to land/vegetation cover; thus, implementation requires in-depth knowledge of the study area. Additional details on radiometric processing can be found in Chapter 2.

Image Transformation and Change Mapping

Classification Scheme/Map Legend

The choice of change detection approach (i.e., categorical comparison and continuous comparison) can profoundly affect the quantitative estimates of forest change (Rogan and Chen, 2004). A problem with many forest cover classification schemes is that the map categories are not always mutually exclusive, which results in class confusion (Gong and Xu, 2003). Class confusion is prevalent in unitemporal forest disturbance mapping (Rogan and Franklin, 2001). For example, locally lower forest biomass caused by an ice storm could be confused with a recently harvested forest stand or senesced pastures with similar spectral properties. GIS data can be invaluable for minimizing class confusion. Bitemporal change detection does not suffer the shortcomings of single-date, postdisturbance methods, but subtle change detection can benefit from the integration of GIS data (Coppin et al., 2004). For example, Rogan et al. (2003) reported that environmental variables such as elevation and slope were selected using a classification tree algorithm when the forest change classification scheme involved nine discrete canopy cover change classes. This situation contrasted sharply with variable selection using a simple change versus no change classification scheme, in which only remotely sensed variables were selected by the classification tree algorithm. Table 6.3 presents examples of studies that have integrated remotely sensed and GIS data for forest change/disturbance mapping and monitoring. Simple topographic variables such as elevation and slope have been used most often in integrative remote sensing-GIS mapping studies.

Land cover and land use data are typically used to perform stratified random sampling for field data collection. Detailed forest inventory information about stand type, structure, and age has been used successfully to map insect damage through the stratification of the calibration data set. These data were used to reduce the variability in the calibration data based on logical decision rules related to host susceptibility and forest structure (see S. E. Franklin and Raske, 1994; S. E. Franklin et al., 2003; Skakun et al., 2003).

Classification Rule

McIver and Friedl (2002) emphasized that all land cover classifications contain elements that reflect analyst expectations. GIS data therefore play a prominent role in providing typicality information as well as ancillary data to guide the choice of decision rule. Slow but continual progress in the integration of spatial analysis software and existing GIS packages has resulted in a growing number of methods from which to choose when formulating inductive models to map forest change (Eastman et al., 2005). Parametric classification algorithms such as maximum likelihood and minimum distance classifiers are available in standard image-processing software. These methods generally produce repeatable and reliable results, but they assume the input data are normally distributed (Carbonell et al., 1983).

Although a standard statistical method in other applications, such as predictive vegetation modeling, generalized linear models (GLMs) have only been applied recently to land cover mapping and monitoring (Morisette et al., 1999; Schwarz and Zimmerman, 2005). GLMs extend classic multiple regression analyses by allowing a less-restrictive form for error distributions (i.e., nonnormal and nonconstant variance functions) (McCullagh and Nelder, 1989). However, GLMs are less exploratory than other more data-driven methods (e.g., classification trees) and require more subjective model specification, requiring that variable transformations and interactions must be explicitly defined a priori.

Nonparametric classification algorithms, such as machine learning, have more recently been applied to forest characterization and change-mapping applications

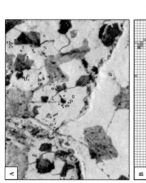

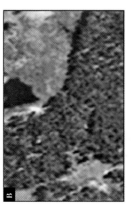

FIGURE 2.1 Illustration of differing information content for three images with differing spatial resolution located near Merritt, British Columbia, Canada. Panel A is an approximately 8 km² area of 30-m spatial resolution Landsat 7 ETM+ multispectral imagery (Path 46/Row 25) collected on August 11, 2001. The 0.05-km² focus area in Panel A is represented in Panels B and C. Panel B is 2.4-m spatial resolution QuickBird multispectral imagery collected on July 17, 2004. Panel C is a digital ortho-image with a spatial resolution of 30 cm collected on August 22, 2003.

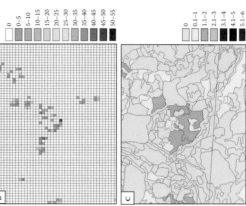

	0
	0–5
	5–10
	10–15
	15–20
	20–25
	25–30
	30–35
	35–40
	40–45
	45–50
	50–55

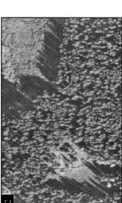

	0
	0.1–1
	1.1–2
	2.1–3
	3.1–4
	4.1–5
	5.1–6

FIGURE 2.4 Illustration of TCT wetness difference image with pixel-level insect infestation locations noted in yellow. Spatial information layers can be developed from the pixel-based infestation locations, such as Panel B, showing the pixel-based disturbance information aggregated as a proportion on a per hectare basis, and Panel C, in which the pixel-based disturbance is summed as an area estimate in hectares on a forest inventory polygon basis.

Clearcut Regeneration Harvest

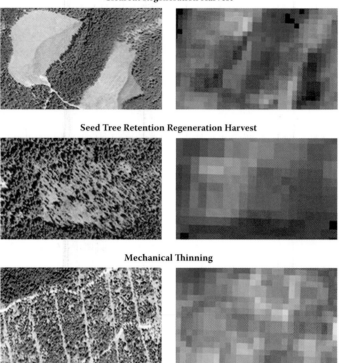

Seed Tree Retention Regeneration Harvest

Mechanical Thinning

FIGURE 3.1 Common harvest practices represented in color orthophotos and tasseled cap-transformed (Crist and Cicone, 1984) Landsat data. Both sets of images were acquired in 2002 and show closed-canopy coniferous forests in central Washington State.

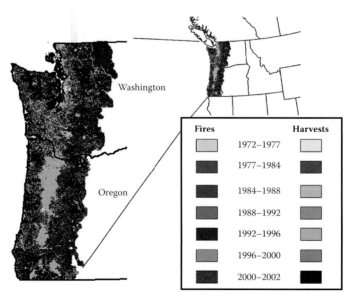

FIGURE 3.4 Map of stand-replacing harvests and fires within the range of the northern spotted owl in Oregon and Washington from 1972 to 2002.

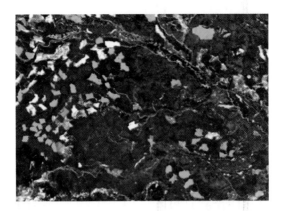

FIGURE 3.6 Three dates of DI as viewed in a typical red-green-blue (RGB) monitor. The first date (1988) is plotted in the red color gun, the second (1992) in the green, and the third (1996) in the blue. Using the assumption that DI is high in disturbed areas, additive color logic can be used to interpret this multitemporal image. Cyan-colored areas are high in both the second and third dates, suggesting a disturbance between the first and second dates. Blue pixels have a high DI only in the third date, indicating the occurrence of a disturbance between the second and third dates. Reddish colors indicate stands disturbed prior to the first date that are becoming revegetated by the second and third dates.

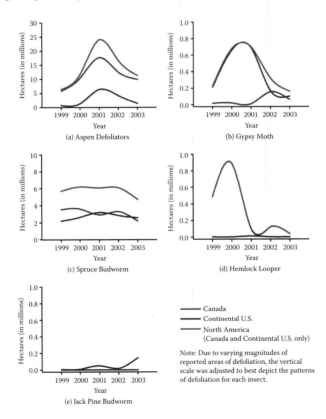

FIGURE 4.1 Patterns of moderate-to-severe insect defoliation from 1999 to 2003 in Canada and the continental United States.

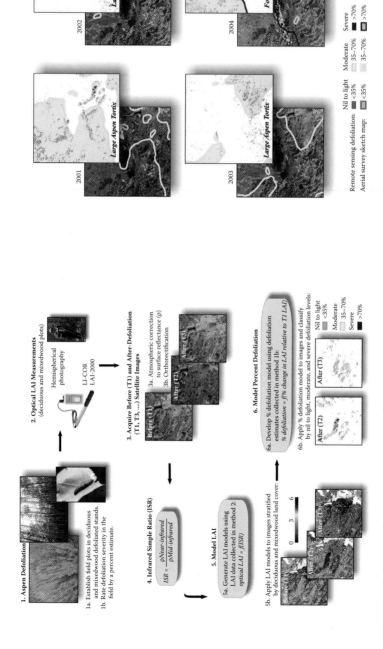

FIGURE 4.2 Method flowchart to generate a time series of aspen defoliation maps based on changes in LAI recorded between pre- and post-defoliation satellite images.

FIGURE 4.5 A subset of the study area illustrating the pattern of insect activity from 2001 to 2004. The aspen defoliation mapped by remote sensing provides a more realistic estimate of the severity and extent of forest damage than the aerial survey sketch maps. Image composites consist of Landsat band 4 as red, band 5 as blue, and band 3 as green using an adaptive stretch enhancement.

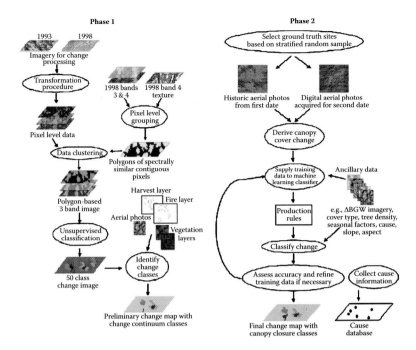

FIGURE 6.3 Overview of LCMMP Phase I and Phase II classification methodology.

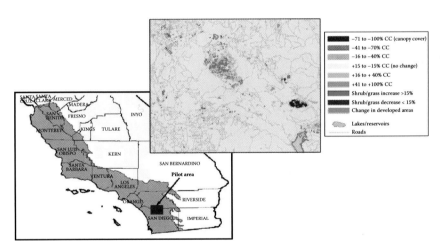

FIGURE 6.4 Case study example of LCMMP map product in Southern California.

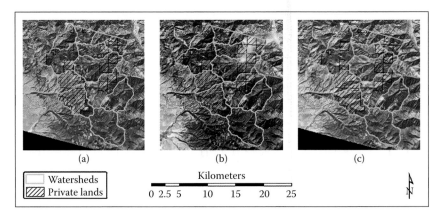

Watersheds
Private lands

Kilometers

0 2.5 5   10   15   20   25

N

FIGURE 8.2 Color infrared composite images of the Cooney Ridge study area (a) 1 year before the wildfire (10 July 2002); (b) during the wildfire (31 August 2003). Note the smoke obscuring the image in the northeastern corner of the burned area; and (c) 1 year after the wildfire (25 September 2004).

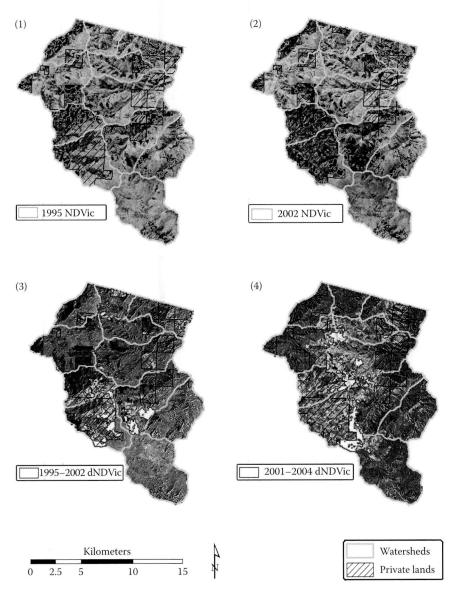

FIGURE 8.4A Stand-replacing disturbance maps: (1) 31 July 1995 NDVIc; (2) 10 July 2002 NDVIc; (3) 31 July 1995 to 10 July 2002 dNDVIc; and (4) 9 September 2001 to 25 September 2004 dNDVIc. The NDVIc-derived polygons indicate patches with minimal forest biomass (Maps 1 and 2), and the dNDVIc-derived polygons indicate patches of stand-replacing disturbance before the 2003 wildfire (Map 3) or as a result of the 2003 wildfire (Map 4). The NDVIc-derived patches are more than two standard deviations below the mean image value, while the dNDVIc-derived patches are more than two standard deviations above the mean image value.

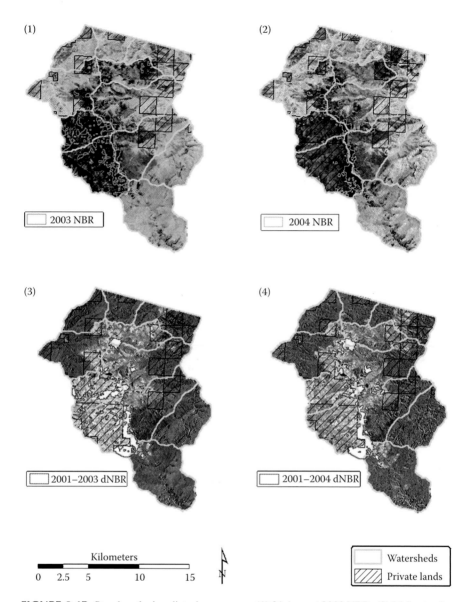

FIGURE 8.4B Stand-replacing disturbance maps: (1) 31 August 2003 NBR; (2) 25 September 2004 NBR; (3) 9 September 2001 to 31 August 2003 dNBR; and (4) 9 September 2001 to 25 September 2004 dNBR. The NBR-derived polygons indicate patches with minimal postfire green vegetation cover (Maps 1 and 2), and the dNBR-derived polygons indicate patches of severe fire-induced tree mortality due to the 2003 wildfire (Maps 3 and 4). The NBR-derived patches are more than two standard deviations below the mean image value, while the dNBR-derived patches are more than two standard deviations above the mean image value.

(Gopal et al., 1999). In addition to often producing better results, they allow for data that are not normally distributed and offer greater ease in incorporating ancillary data (Friedl et al., 1999; Skidmore and Turner, 1988). MLAs refer to the application of induction algorithms that analyze information, recognize patterns, and improve prediction accuracy through automated, repeated learning from training data (Carbonell et al., 1983; Malerba et al., 2001). There is a large body of research that demonstrates the abilities of machine learning techniques, particularly classification trees and artificial neural networks, to deal effectively with tasks involving data with high dimensionality because of their ability to reduce computational demands significantly for nonlinear data distributions (Gahegan, 2003). As a result, MLAs have gained acceptance in the context of large-area forest mapping (Friedl et al., 1999; Hansen et al., 1996) given the need for automated, objective, and reproducible classification methods that can handle very large volumes of data across coarse spatial scales (Borak and Strahler, 1999; Gopal et al., 1999; Hansen et al., 2000; Hansen and Reed, 2000; Muchoney and Williamson, 2001). Classification and regression trees have the potential to serve as rule generators for complex forest-monitoring tasks because mapping decisions are transparent and explicit (Rogan et al., 2003). In addition, decision trees can provide a "first cut" rule set as input to an expert system and act as a bridge between automation and expert knowledge.

Expert system (also referred to as knowledge- or rule-based) classifiers are a recently explored alternative to conventional supervised classification. Expert systems typically relate classes to properties through a series of rules representing conditional statements and are favored for complex mapping tasks (X. Huang and Jensen, 1997). Expert system approaches have several important advantages related to their facility for incorporating GIS data in the rule-making (knowledge-generating) process. Unlike many statistical methods, expert systems do not have stringent requirements about data distribution and independence (Quinlan, 1993). Expert systems have been used to incorporate GIS data in several land cover mapping applications (Comber et al., 2004; X. Huang and Jensen, 1997; Levien et al., 1999; Raclot et al., 2005; Rogan et al., 2003). Ehlers et al. (2003) developed an automated procedure for incorporating GIS layers, elevation, and multispectral image data for fine spatial resolution biotype mapping.

Object-based or per-field approaches to forest cover mapping and monitoring appear to be an emerging theme in the GIS science literature (Raza and Kaiz, 2001). Despite widespread use, pixel-based methods for mapping generally do not make use of the spatial and geometric properties of the data (Wulder, 1998). Object-based classification methods allow for the incorporation of contextual information in the mapping process. This type of application enables the segmentation of multispectral imagery into meaningful homogeneous objects, or regions, based on neighboring pixel spectral and spatial values. Although operational examples are rare, case studies involving a variety of remotely sensed data types are becoming more common (Lamar et al., 2005). Wulder et al. (2004) used segmented Landsat-7 Enhanced Thematic Mapper Plus (ETM+) data to estimate stand ages of regenerating lodgepole pine forest stands. Segmentation aided the removal or masking of pixels on the periphery of clearcuts that consisted of intact trees. Segmented polygons were shown to represent more accurately estimated stand age than forest inventory polygons.

Also, Hinton (1999) integrated airborne synthetic aperture radar and vector data representing forest stands of a single species to minimize within-class confusion and produced maps with 8% higher accuracy than a per-pixel method. Spatial errors in the vector data initially resulted in a reduction in map accuracy prior to some additional processing. One substantial drawback of object-based methods is that land cover change may only be detected if a substantial proportion of the object is modified. For example, when the unit of observation is a forest stand, multitemporal changes do not always follow stand delineations (Varjo, 1997). The question of whether stand delineations should be based on acquired images or existing GIS coverages is subject to further investigation.

VALIDATION AND CHANGE ANALYSIS

An in-depth understanding of the processes of forest change/disturbance is predicated on the ability to monitor forests accurately over several decades (Lambin, 1999). A need exists for operational methods to assess the quality of large-area change maps. Unfortunately, well-established accuracy assessment methods (i.e., using an independent sample of validation data) that are commonly used at local scales are often not practical at coarse scales. Validation can be based on existing maps. Siqueira et al. (2000) validated a land cover map of the Amazon, based on Japanese Earth Resources Satellite (JERS-1) data, using a combination of physiographic/climate-based vegetation maps, local vegetation maps, and Advanced Very High Resolution Radiometer (AVHRR)-based land cover maps to estimate 14 vegetation classes with an accuracy of 78%. High-quality local-scale maps can be used for large-area validation (Siqueira et al., 2003). Stoms (1996) proposed the use of "maplets" for validating large data sets — maps from local and state agencies for specific sites (e.g., a state park or a project area). While promising, these fine spatial resolution products require careful processing and preparation if they are to be used to validate coarser resolution products acquired by different sensors for different operational mapping needs (Trietz and Rogan, 2004). For example, uncertainty can be introduced into validation results as a consequence of differences between the classification schemes of each map and potential geolocation errors in both products (Fuller et al., 2003).

Efforts to compare different land cover products hinge on interoperability between remotely sensed and GIS data sets. Inconsistencies between spatial resolutions and land cover classification systems inhibit comparison and generalization between large-area regional monitoring systems and global monitoring systems. Recent work in GIS science has begun to reconceptualize the basis of land cover classification systems by defining classes with formal parameterizations (Ahlqvist, 2004). Land cover classes are viewed as semantic concepts that can be defined by quantitative parameters, such as percentage cover of tree crowns or texture indices. Each class is defined by a collection of fuzzy set membership functions for a specified number of continuous variables that describe the class. Salience weights are applied to specify the importance of each variable to the definition of the land cover class. To address the complexities of having different spatial resolutions in map comparison, Pontius (2002) presented new statistical methods to partition effects of quantity and location in a comparison of categorical maps at multiple spatial resolutions.

CURRENT LIMITATIONS OF FOREST CHANGE
DETECTION AND MAPPING STUDIES

Data recorded by remote sensing instruments are valuable for providing information on forest cover conversion and modification but are not always a consistent indicator of discrete change events (Loveland et al., 2002). The detectability and accurate characterization of forest disturbance using remotely sensed data are influenced by the type of disturbance, the magnitude and duration of the modified signal, and natural variability (species/stand/landscape). These factors can often result in high errors of omission and commission in forest change maps. Indeed, map accuracy in land cover change research is typically 15–20% lower than that found in single-date land cover scenarios (Rogan et al., 2003). Figure 6.2 presents the conceptual trade-offs that exist in a forest disturbance mapping scenario. Trade-off considerations of typicality, data characteristics, and the classification/mapping rule become increasingly problematic as landscape heterogeneity increases because the spectral variation caused by forest decline often overlaps with spectral variation caused by topography, species composition, and stand structure (S. E. Franklin and Raske, 1994). For example, large-area mapping/monitoring is especially difficult because any landscape homogeneity at small spatial extents (e.g., a single Landsat image) can transform into heterogeneity at larger extents (e.g., a mosaic of ten Landsat images) (Wulder et al., 2004).

OMISSION ERRORS

Problem

Anthropogenic disturbances such as forest conversion to agriculture or urban land use are typically mapped with replicable levels of map accuracy (Seto et al., 2002). When the forest disturbance does not cause an acute alteration in the spectroradiometric or textural properties of the landscape, excessive omission errors are common. For example, Olsson (1995) could not reliably map canopy cover decrease (less than 20–25%) caused by forest thinning using Landsat-5 TM data in a boreal forest because damage did not result in a near-infrared reflectance decrease in excess of 0.015. Similarly, Souza and Barreto (2000) could not reliably detect/map the locations of selective harvest sites in tropical forest, using Landsat-7 ETM+ data, despite having access to detailed field data on timber extraction.

When researchers seek to derive ordinal-scale disturbance information at fine levels of thematic detail (i.e., high, medium, low), detection accuracy is often less reliable for low-impact categories (Coops et al., Chapter 2, this volume; Rogan et al., 2003). Dichotomous categories of forest change/no change can usually be mapped using Landsat-like data with accuracies on the order of 80–90% overall accuracy (Varjo, 1997). Detection of ordinal change/disturbance categories is a more difficult task (Heikkonen, 2004), requiring more stringent requirements for calibrated and corrected satellite data to remove noise. For example, Rogan and Franklin (2001) reported that light wildfire burn severity areas were less reliably mapped than *severe* areas in chaparral woodland because of spectral confusion with unburned vegetation patches. Ekstrand (1990) reported that reflectance in Norway spruce decreased as

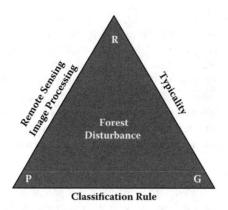

FIGURE 6.2 Diagram illustrating the trade-offs that exist in forest disturbance/change mapping where R = Reality; P = Precision; G = Generality. Typicality (information classes) and spectral classes (remote sensing) are rarely in complete agreement with each other (i.e., R G). Classification (decision) rules make suboptimal choices in the presence of image noise or suboptimal sensor/data/scene models (i.e., P R). Generalization of well-modeled decision rules is rarely possible over highly heterogeneous landscapes/regions/biomes (i.e., P G).

needle loss increased from 10 to 40% due to tissue damage and pigment alterations. Variation in composition and density of forest stands was cited as the cause of low accuracy levels because intercanopy shadowing affected the ability to discriminate between low and moderate defoliation levels and caused spectral differences between areas of similar defoliation conditions. Gemmell and Varjo (1999) reported problems in detecting different levels of timber harvest in a boreal forest caused by variation in tree species composition, type, and understory vegetation. Ciesla et al. (1989) reported confusion between moderate levels of gypsy moth damage and conifer plantations on shaded slopes.

Change detection approaches used to classify the cause of disturbance are problematic in that classes are not mutually excusive and totally exhaustive (Coppin and Bauer, 1996). Further, several authors have found that partial-cut classes often have higher omission errors than clearcuts (Heikkonen, 2004; E. H. Wilson and Sader, 2002). Higher temporal-spectral variation in peat soils versus mineral soils has also caused confusion in mapping timber harvest (Varjo, 1997). While the radiation ecology of forest understory has been cited as a source of confusion in detecting overstory damage, subcanopy disturbances such as surface wildfires, brush clearing, and floods are largely undetectable because the "disturbance signal" is blocked by the overstory canopy (Rogan and Franklin, 2001). Other confounding influences on the detectability of forest change include the rates of regeneration and recovery, typically determined by climatic factors, plant adaptation to disturbance, and mitigation of impact by resource managers (Coppin et al., 2004).

Solution

Used judiciously, GIS data have the potential to mitigate some of the current limitations of forest change mapping associated with map omission errors. In the first instance, information about forest cover or soil type could be used to help control for the intrinsic landscape variability that can prevent adequate detection of subtle disturbances (Heikkonen, 2004). GIS data can also provide cues and clues for detection of *cryptic* target features, such as selective harvest in tropical forest, based on information from other scene objects such as the use of roads and log landings (Nepstad et al., 1999; Souza et al., 2005). Second, GIS data can be used directly to provide additional predictive power to a classification algorithm (Eastman et al., 2005). For example, using remotely sensed data alone, one cannot detect a burn scar that occurred underneath a closed forest canopy. Burn detection accuracy, however, may increase if a slope variable is added to the analysis when training data also represent burn locations on certain slope intervals (Rogan et al., 2003). GIS data can thus be used as environmental input variables potentially to serve to fill image data "gaps." From a temporal perspective, Brivio et al. (2002) found digital topographic data useful to overcome the limitation caused by the time lag between the peak of a flooding event and European Remote-Sensing Satellite (ERS-1) synthetic aperture radar satellite overpass to map flood damage.

COMMISSION ERRORS

Problem

Even if the signal of a forest disturbance is strong enough to overcome species-stand variation to be detectable, it can be confused with landscape features having similar spectral properties, resulting in map commission errors. Commission errors are most prevalent in unitemporal change detection studies, in which only postdisturbance images are used, and analysts do not have the benefit of multispectral "from-to" information, common in bitemporal change detection (Coppin et al., 2004). Burn scars are typically confused with asphalt roads and deep, clear water bodies in most environments (e.g., Chuvieco and Congalton, 1988); secondary growth is often mislabeled as pasture (and vice versa) in disturbed tropical forest (Powell et al., 2004); recent timber clearcuts are occasionally misclassified as senesced meadows in temperate forests (Levien et al., 1999); and heavy levels of gypsy moth damage are confused with fallow fields and talus slopes (Ciesla et al., 1989). Such problems of signature separability typically bias severe or high damage/impact disturbance categories, resulting in "overclassification" in ordinal-scale disturbance maps.

The presence of other disturbance types (current or past) is a second source of commission errors in forest change maps. In chronically modified landscapes, different types of natural and anthropogenic disturbances may spatially coexist and interact over time (Attiwill, 1994; Kittredge et al., 2003; Souza et al., 2005). For instance, the presence of old burn scars in a study area can be misclassified as light burn (Key and Benson, 2002), and ice storm damage to canopy can be confused with selective harvest removal. Further, it has been demonstrated that one type of disturbance can influence the proclivity of a forest stand to future disturbance. Macomber and Woodcock (1994) and Collins and Woodcock (1996) studied the impacts of drought on insect pest mortality in temperate forests. Also, Lindemann and Baker (2002) found forest blowdown sensitive to physical factors such as wind exposure, aspect, elevation, and forest cover type, while others have cited the significant influence of previous clearcutting on windthrow (see Coates, 1997; Huggard et al., 1999).

Solution

To mitigate commission errors in disturbance mapping, GIS data can be used to create spatial masks of the landscape features most often confused with the disturbance type of interest (Levien et al., 1999). Objects with low optical reflectance such as water and roads as well as topographic shadows can often be confused with severely disturbed areas such as severe burns. Also, early stage secondary forest growth is often indistinguishable from shrub and scrub vegetation (see Powell et al., 2004). Where multiple disturbances are present or have a high probability of presence, GIS data representing previous disturbance events could be used to reduce the potential confusion. For example, White et al. (2005) applied an "exclusionary" mask generated from the locations of logged sites, water bodies, and cloud cover to IKONOS-2 images to reduce variability in mapping mountain pine beetle outbreak.

SELECTED APPLICATIONS

BURN MAPPING

Detailed burn scar mapping is one of the most challenging applications of remotely sensed data and remote sensing technology (Rogan and Yool, 2001). In addition to the removal of vegetation and exposure of soil, the aftermath of combustion adds new features to a remote sensing scene — charcoal and ash. Forest environments burn with varying intensities (i.e., energy released per unit length of flame front, per unit time), depending on fuel type, fuel load, fuel moisture, and topographic constraints (i.e., slope and aspect) (Pyne et al., 1996). Variation in fire intensity yields variations in burn severity, ranging widely from partial consumption of vegetation cover with little soil exposure or char/ash deposition, to complete consumption of vegetation cover with high soil exposure and char/ash deposition (Pyne et al., 1996; Yool et al., 1985). Consequently, the cumulative effect of a burn is often a heterogeneous mix of remote sensing image scene elements associated with burn severity or damage to soil and vegetation (Clark and Bobbe, Chapter 5, this volume; Rogan and Franklin, 2001; A.M.S. Smith et al., 2005).

Several problems make burn severity monitoring difficult using satellite imagery (i.e., at coarse [AVHRR] or finer [Landsat-5 TM] spatial scales). The most commonly reported problem is that burned vegetation patches are often confused spectrally with nonvegetated surfaces with similar spectral signatures (i.e., asphalt roads, deep water bodies). However, the effects of topography and smoke plumes confound these factors. Topographically induced shade caused by illumination differences can create spectral confusion between shaded unburned vegetated patches, shaded nonvegetated patches, and burned patches (Chuvieco and Congalton, 1988). Over large areas, vegetation diversity becomes problematic as it is difficult to assign a label of high, medium, and low damage when vegetation diversity (lifeform) is also spatially variable (Rogan and Franklin, 2001). Thus, stratification by soil or vegetation would be useful (Vigilante et al., 2004). Several applications to map burn scars have successfully incorporated GIS and remotely sensed data (Chuvieco and Congalton, 1988; White et al., 1996). Medler and Yool (1997) combined composite terrain and Landsat-5 TM imagery in a supervised classification to map wildfire mortality. Error matrices indicated that this amalgam of satellite and ancillary data provided a 40% improvement in accuracy compared to TM data alone. DEMs are not always useful, however, when burn management/containment strategies are contrary to theoretical fire behavior models (Rogan and Yool, 2001).

PEST INFESTATION

Many methods used for detecting insect defoliation were originally developed to detect forest damage related to air pollution in European forests (Waldsterben) in remotely sensed imagery (Herrmann et al., 1988). Factors contributing to insect pest infestation include drought stress (Collins and Woodcock, 1996); high stand density; species composition, age, elevation, aspect, vigor (S. E. Franklin, 2001); and soil type (Bonneau et al., 1999). Compared to burn mapping, there are fewer operational examples of pest damage mapping using medium spatial resolution data. This is due to the high natural variability in forests affected by pests and the relatively light influence of pest damage on the spectral response of medium-resolution/broadband sensors (S. E. Franklin et al., 2003). For example, Nelson (1983) reported that a moderate pest defoliation category could not be accurately delineated using Landsat MSS data as it was usually confused with the reflectance variability of healthy forest. This problem was also reported by Joria et al. (1991) using both Landsat-5 TM and SPOT-2 data.

To address the previously stated challenges, Williams and Nelson (1986) developed techniques using a Landsat-5 TM Band 5/7 (mid infrared) ratio to delineate and assess forest damage due to defoliating insects; they reported 90% overall accuracy for delineating insect-damaged and healthy forest. The use of a nonforest mask reduced classification confusion with nondefoliated areas in the scene that displayed similar reflectance to defoliated canopy. Rohde and Moore (1974) analyzed single- and multidate Landsat-1 data to detect the impact of gypsy moth. In this early study, confusion of sites with defoliation with agricultural land use, or open-face mining areas in postinfestation imagery only, was minimized using the multidate images. S. E. Franklin et al. (2003) and Skakun et al. (2003) examined mountain pine beetle red attack damage in lodgepole pine stands in British Columbia. Overall map accuracies of 73–78%, using postinfestation Landsat-5 TM data, were facilitated through the stratification of the calibration data set using polygonal forest inventory data. These data were used to reduce the variability in the calibration data based on logical decision rules related to host susceptibility and forest structure. This stratification technique was applied previously to improve classification results of spruce budworm defoliation in western Newfoundland (S. E. Franklin and Raske, 1994).

Wulder et al. (2005) used a polygon decomposition approach to integrate different sources of data (field data, aerial surveys, Landsat images) within a GIS to examine the impacts of mountain pine beetle. Polygon decomposition was implemented by populating forest inventory polygons with the proportion (in percentage) and area (in hectares) of damaged pixels that had been generated from Landsat-7 ETM+ data. Analysis of the combined data revealed that stands of high pine component in the age category 121 to 140 years, with diameter breast heights above 25 cm, and with 66 to 75% crown closures were most susceptible to beetle attack. Younger balsam fir (*Abies balsamea*) stands are more susceptible to spruce budworm defoliation than mature stands. This makes satellite-derived age maps suitable for mapping a determining factor of insect population levels useful in predicting future outbreaks (Luther et al., 1997). Finally, Radeloff et al. (1999) excluded timber clearcut areas and masked pure stands to detect jack pine budworm defoliation.

ICE STORM DAMAGE

Remote sensing applications in ice storm damage mapping were rare in the remote sensing literature until the 1998 ice storm event, which affected large portions of northern New England in the United States and southern Quebec and eastern Ontario in Canada (Irland, 1998). Ice storm damage to forest canopy occurs at different spatial and temporal scales and results in bending, branch loss, and topping. Damage is often related to canopy architecture, tree size, age, health, and the mechanical properties of the wood itself (e.g., elasticity and rigidity) (Pellikka et al., 2000). Physiographic factors such as elevation and slope influence the depth and duration of ice accumulation (Irland, 1998). Olthof et al. (2004) found that a neural network classifier produced damage maps with higher accuracies than the conventional parametric classifiers when ancillary environmental variables (ice accumulation, elevation, slope, aspect, distance from forest edge) were incorporated into the classification process. Classification accuracy improved from light, medium, to heavy damage categories (19.5%, 44.4%, 77.3%, respectively). Overall damage classification accuracy was approximately 65%. Ice storm damage is indicative of the complexity found when seeking trends related to the influence of environmental factors resulting from a particular disturbance event (e.g., elevation, aspect, slope, and forest type). In reviewing the increasing body of literature on the topic, environmental variables such as "distance to forest edge" played a significant role in damage prediction in some, but not all, studies.

TIMBER HARVEST

Timber clearcut detection and monitoring appears to be one of the most successful and reliable applications of remotely sensed data in forest disturbance mapping. Franklin et al. (2000) presented a comprehensive examination of harvest-related change over 15 years using Landsat-5 TM data in the Fundy Model Forest, New Brunswick (Canada). Multitemporal change thresholds (based on Kauth-Thomas wetness) were calculated based on spatial information concerning areas disturbed by clearcutting, partial harvesting, and silvicultural treatments. Further, GIS inventory data were used to mask all nonforest areas for final forest change mapping. Gemmell and Varjo (1999) reported problems in detecting different levels of timber harvest in a boreal forest caused by variation in tree species composition, type, and understory vegetation. In addition, when the unit of observation is a forest stand, multitemporal changes do not always follow stand delineations (Varjo, 1997).

Heikkonen (2004) prestratified a boreal forest area based on mineral soil type only to reduce the influence of soil background variation timber harvest mapping. Masks are often applied to eliminate "irrelevant areas", such as water bodies and extreme slopes, from analysis (Wilson and Sader, 2002, p. 7). Saksa et al. (2003) promoted the use of predelineated segments or pixel blocks for image differencing to decrease the number of misinterpreted areas in a study. In this work, a digital forest mask was considered "crucial" to operational applications. Nilson et al. (2001) stated that thinning in boreal forest could result in the appearance of bare soil, cutting waste, and subcanopy vegetation. Heikkonen (2004) found a moderate harvest category (thinning and preparatory cut) least accurate compared to no change and considerable change categories. Selective logging is becoming a major form of disturbance in tropical forests (Cochrane and Souza, 1998). These modifications are often difficult to detect (Coops et al., Chapter 2, this volume). Conway et al. (1996) improved the detection of selectively logged areas using expert knowledge of the topography and soil disturbance patterns of logged tropical forests.

CASE STUDY: THE CALIFORNIA LAND COVER MAPPING AND MONITORING PROGRAM

To address the growing threat to forest and shrubland sustainability caused by rapid and widespread land cover change in California, the U.S. Forest Service and the California Department of Forestry and Fire Protection are collaborating in the state-wide Land Cover Mapping and Monitoring Program (LCMMP) to improve the quality and capability of monitoring data and to minimize costs for statewide land cover monitoring (Levien et al., 1999). The long-term goals of the LCMMP are to develop a baseline to monitor the amount and extent of forest and rangeland resources, to track forest health trends, and to examine the effectiveness of existing environmental policies. Monitoring data created by the LCMMP quantifies changes to forests, shrublands, and urban areas across 70% of California and provides necessary information for regional assessment across jurisdictional boundaries (Levien et al., 1999). A key advantage of this cooperative program is that monitoring information provides a single consistent source of current landscape-level and site-specific change to the U.S. Forest Service and California Department of Forestry and Fire Protection as well as other interested federal agencies. The LCMMP maps and monitors land cover according to the boundaries of 20 or more Landsat scenes and ecological subsections from the National Hierarchical Framework of Ecological Units (Bailey, 1983). The total area includes approximately 2 million Ha of National Forest Service lands.

The data-processing flow of the LCMMP is presented in Figure 6.3. The LCMMP uses Landsat-5 TM and Landsat-7 ETM+ satellite imagery within five-year monitoring periods. Changes in forest, shrub, and grassland cover types are the primary focus of this program, but changes in urban/suburban areas are also mapped (Table 6.4). These change maps are required for regional interagency land manage-

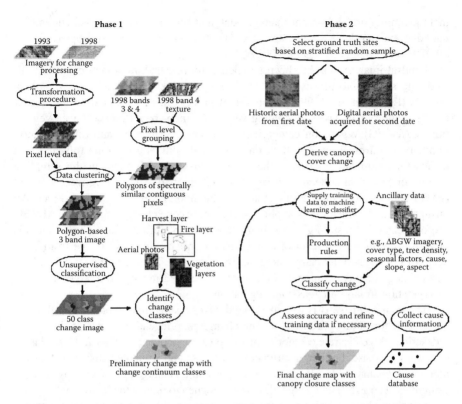

FIGURE 6.3 (See color insert following page 146.) Overview of LCMMP Phase I and Phase II classification methodology.

TABLE 6.4
Classification Schemes for LCMMP Phase I and Phase II Land Cover Change Maps

Phase I change classes	Phase II change classes
Large decrease in vegetation	−71 to −100% canopy change
Moderate decrease in vegetation	−41 to −70% canopy change
Small decrease in vegetation	−16 to −40% canopy change
	Shrub/grass decrease > 15%
Little or no change	±15% canopy change
Moderate increase in vegetation	+16 to +40% canopy change
Large increase in vegetation	+41 to 100% canopy change
	Shrub/grass increase > 15%
Nonvegetation change	Change in developed areas

ment planning, fire and timber management, and species habitat assessment and for updating existing land cover maps at a low cost per unit area cost (approximately $0.01/Ha) (Levien et al., 1999, 2002).

Landsat imagery that has been geometrically rectified, radiometrically normalized, and subset into processing areas is ready for input into the change-mapping process (Levien et al., 1999). A concurrent process involves preparing and mosaicking ancillary data layers, including vegetation maps based on California Vegetation (CALVEG) vegetation categories, fire history perimeters, and timber plantation/harvest information. Ancillary data are used both as a masking tool and as a means for stratification to label the change classes and implement the sampling design for field data collection. Image features are then extracted from the Landsat (E)TM data using texture (Band 4) and multitemporal Kauth-Thomas routines. An unsupervised classification is applied to each per-scene change image by CALVEG lifeform category, resulting in 50 change classes per lifeform. Each change class is labeled according to its level of change based on a gradient of change from large decreases in vegetation to large increases in vegetation. The final product from Phase I is a change map containing a gradient of classes that ranges from large decreases in vegetation to large increases in vegetation.

The goal of Phase II is to make a land cover change map representing discrete changes in forest and shrub cover. The change map legend is shown in Table 6.4; it describes three discrete categories of forest canopy cover decrease and two classes of canopy increase. Further, a shrub cover increase and shrub decrease class is used, along with change in developed (urban) areas and no change (±15% canopy change) categories. The ±15% no change class was designed to reduce the confusion between phenological changes and postdisturbance increase classes by allowing for minor increases/decreases in vegetation abundance. Ground reference data are obtained by estimating forest canopy cover change within a five-year time frame using two sets of aerial photographs and *in situ* information. To calibrate canopy cover estimates from air photos, photo-interpreted canopy cover measurements are compared with those measured in the field. The result of this analysis is a calibration/validation data set portraying classes of canopy cover change, which are then used in the change-mapping process.

An example of the Phase II change map for a pilot study region in southern California is shown in Figure 6.4. A general map accuracy assessment is performed on the land cover change maps. Final products from Phase II include the land cover change map derived from the classifier featuring discrete canopy cover change classes and the GIS database identifying the locations of vegetation change with cause information for coniferous forest, hardwood rangeland, shrub cover, and urban areas. This product is then made available to various national resource agencies for ecosystem management activities (Levien et al., 1999). Related products can be found at: http://www.fs.fed.us/r5/spf/about/fhp-change.shtml.

CONCLUSIONS AND FUTURE DIRECTIONS

"The mapping and measurements of small to medium scale changes over large areas requires levels of precision in mapping which are near impossible to achieve with

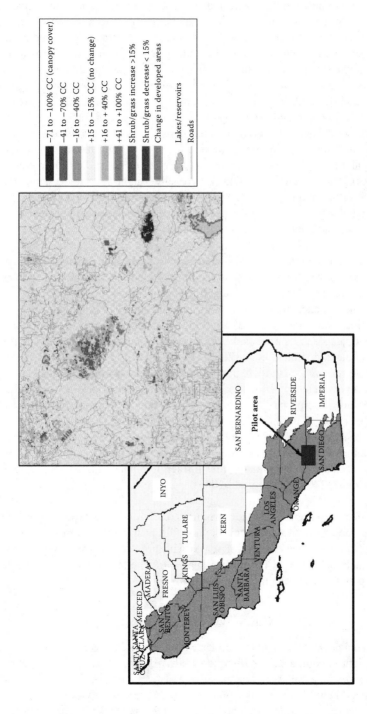

FIGURE 6.4 (See color insert following page 146.) Case study example of LCMMP map product in Southern California.

satellite image classifications alone" (Fuller et al., 2003, p. 252). In response to the potentially confounding effects forest disturbances have on single-date mapping applications and the confusion caused by landscape heterogeneity on disturbance mapping applications, recent research has focused on strategies to reduce map confusion using new methods or additional data (Rogan et al., 2003). Advances in GISystems and GIS data availability, quality, and type in combination with advances in GIScience research can potentially mitigate the current challenges of large-area monitoring and detailed investigations of subtle forest modifications — two chief impediments to in-depth understanding of the scale and pace of forest change. Smith et al. (2003) stated that digital remotely sensed imagery is soon likely to be a standard instrument in the repertoire of the professional forest manager because the nexus of technology and need has finally occurred. Ustin et al. (2004, p. 689), however, held that "after 30 years of remote sensing, we are still struggling to understand how to interpret the information content in images." To integrate better GIS and remotely sensed data and technology with the needs of forest management, we present some important research themes for the near future:

1. Future developments should include expert systems that make better use of multisensor approaches and context-based interpretation schemes (Davis et al., 1991; Fuller et al., 2003).
2. GIS intelligence (e.g., object and analysis models) should be used to automate the forest change/disturbance classification process. In return, GIS objects can be extracted from a remotely sensed image to update the GIS database (Ehlers et al., 2003).
3. Single research methodologies do not suffice for a complete analysis of forest cover change. Instead, a sequence of methodologies is needed that integrates disciplinary components over a range of spatial and temporal scales (S. E. Franklin and Wulder, 2002).
4. Representation of land cover as continuous fields of various biophysical variables for accurate detection of forest degradation (Lambin, 1999).
5. Increased data integration requires further investigation into data accuracy (Ahlqvist, 2004; Fuller et al., 2003).

While there are some outstanding issues to address, data products developed through the integration of remote sensing and GIS enable the capture and representation of disturbance over a range of scales with predictable results. These disturbance products are suitable for further analysis to better inform landscape level dynamics, patterns, and resultant implications.

ACKNOWLEDGMENTS

We wish to thank Dr. Mike Wulder and Dr. Steven Franklin, the editors of this book, for their insightful comments and direction while we were preparing this chapter. We also acknowledge the collaboration and advice of Lisa Fischer (U.S. Forest Service, California) and helpful comments from Dr. Tim Warner (West Virginia University), Tim Currie, Steven McCauley, Zachary Christman, Rima Wahab-Twibell (Clark University), and Massachusetts Forest Monitoring Project (MAFoMP) student researchers (see: http://www.clarku.edu/departments/hero/research/forestchange.htm).

REFERENCES

Ahlqvist, O. (2004). A parameterized representation of uncertain conceptual spaces. *Transactions in GIS, 8*, 493–514.

Aspinall, R. (2002). A land-cover data infrastructure for measurement, modeling, and analysis of land-cover change dynamics. *Photogrammetric Engineering and Remote Sensing, 68*, 1101–1105.

Attiwill, P.M. (1994). The disturbance of forest ecosystems: the ecological basis for conservative management. *Forest Ecology and Management, 63*, 247–300.

Austin, M.P. and Smith, T.M. (1989). A new model for the continuum concept. *Vegetatio, 83*, 35–47.

Bailey, R.G. (1983). Delineation of ecosystem regions. *Environmental Management, 7*, 365–373.

Baker, W.L. (1989). A review of models of landscape change. *Landscape Ecology, 2*, 111–133.

Bergen, K.M., Brown, D.G., Rutherford, J.F., and Gustafson, E.J. (2005). Change detection with heterogeneous data using ecoregional stratification, statistical summaries and a land allocation algorithm. *Remote Sensing of Environment, 97*, 434–446.

Bolstad, P.V. and Stowe, T. (1994). An evaluation of DEM accuracy: elevation, slope, and aspect. *Photogrammetric Engineering and Remote Sensing, 60*, 1327–1332.

Bonneau, L.R., Shields, K.S., and Civco, D.L. (1999). Using satellite images to classify and analyze the health of hemlock forests infested by the hemlock woolly adelgid. *Biological Invasions, 1*, 255–267.

Borak, J. and Strahler, A. (1999). Feature selection and land cover classification of a MODIS-like data set for a semi-arid environment. *International Journal of Remote Sensing, 20*, 919–938.

Bricker, O.P. and Ruggiero, M.A. (1998). Toward a national program for monitoring environmental resources. *Ecological Applications, 8*, 326–329.

Brivio, P.A., Colombo, R., Maggi, M., and Tomasoni, R. (2002) Integration of remote sensing data and GIS for accurate mapping of flooded areas. *International Journal of Remote Sensing, 23*, 429–441.

Brown, J.F., Loveland, T., Merchant, J., Reed, B., and Ohlen, D. (1993). Using multisource data in global land characterization: Concepts, requirements, and methods. *Photogrammetric Engineering and Remote Sensing, 59*, 977–987.

Cablk, M.E., Kjerfve, B., Michener, W.K., and Jensen, J.R. (1994). Impacts of Hurricane Hugo on a coastal forest: assessment using Landsat TM data. *Geocarto International, 2*, 15–24.

Carbonell, J.G., Michalski, R.S., and Mitchell, T.M. (Eds.). (1983). *An Overview of Machine Learning: An Artificial Intelligence Approach*, Vol. 1. Tioga Publishing Co., Palo Alto, CA. 122p.

Carpenter, G.A., Gopal, S., Macomber, S., Martens, S., Woodcock, C.E., and Franklin, J. (1999). A neural network method for efficient vegetation mapping. *Remote Sensing of Environment, 70*, 326–338.

Chafer, C.J., Noonan, M., and Mcnaught, E. (2004). The post-fire measurement of fire severity and intensity in the Christmas 2001 Sydney wildfires. International Journal of Wildland Fire, 13(2): 227–240.

Chen, K. (2002). An approach to linking remotely sensed data and areal census data. *International Journal of Remote Sensing, 23*, 37–48.

Chuvieco, E. and Congalton, R.G. (1988). Mapping and inventory of forest fires from digital processing of TM data. *Geocarto International, 4*, 41–53.

Cibula, W. and Nyquist, M. (1987). Use of topographic and climatological models in a geographic data base to improve Landsat MSS classification for Olympic National Park. *Photogrammetric Engineering and Remote Sensing, 53*, 67–75.

Ciesla, W.M., Dull, C.W., and Acciavatti, R.E. (1989). Interpretation of SPOT-1 color composites for mapping defoliation of hardwood forests by gypsy moth. *Photogrammetric Engineering and Remote Sensing, 55*, 1465–1470.

Coates, D. (1997). Windthrow damage 2 years after partial cutting at the Date Creek Silvicultural Systems Study in the interior cedar-hemlock forests of northwestern British Columbia. *Canadian Journal of Forest Research, 27*, 1695–1701.

Cochrane, M.A. and Souza, C.M. (1998). Linear mixing model classification of burned forests in the eastern Amazon. *International Journal of Remote Sensing, 19*, 3433–3440.

Collins, J.B. and Woodcock, C.E. (1996). An assessment of several linear change detection techniques for mapping forest mortality using multitemporal Landsat TM data. *Remote Sensing of Environment, 56*, 66–77.

Comber, A.J., Law, A.N., and Lishman, J. (2004). Application of knowledge for automated land cover change monitoring. *International Journal of Remote Sensing, 25*, 3177–3192.

Congalton, R.G., Birch, K., Jones, R., and Schriever, J. (2002). Evaluating remotely sensed techniques for mapping riparian vegetation. *Computers and Electronics in Agriculture, 37*, 113–126.

Conway, J., Eva, H., and D'Souza, G. (1996). Comparison of the detection of deforested areas using the ERS-1 ATSR and the NOAA-11 AVHRR with reference to ERS-1 SAR data: A case study in the Brazillian Amazon. *International Journal of Remote Sensing, 17*, 3419–3440.

Coppin, P., Jonckheere, I., Nackaerts, K., Muys, B., and Lambin, E. (2004). Digital change detection methods in ecosystem monitoring: A review. *International Journal of Remote Sensing, 25*, 1565–1596.

Coppin, P.R. and Bauer, M.E. (1996). Digital change detection in forest ecosystems with remote sensing imagery. *Remote Sensing Reviews, 13*, 207–234.

Coulter, L., Stow, D., Hope, A., O'Leary, J., Turner, D., Longmire, P., Peterson, S., and Kaiser, J. (2000). Comparison of high resolution imagery for efficient generation of GIS vegetation layers. *Photogrammetric Engineering and Remote Sensing, 66*, 1329–1335.

Davis, F.W., Quattrochi, D.A., Ridd, M.K., Lam, N.S.-N., Walsh, S.J., Michaelsen, J.C., Franklin, J., Stow, D.A., Johannsen, C.J., and Johnston, C.A. (1991). Environmental analysis using integrated GIS and remotely sensed data: some research needs and priorities. *Photogrammetric Engineering and Remote Sensing, 57*, 689–697.

Dobson, J.E. (1993). A conceptual framework for integrating remote sensing, GIS, and geography. *Photogrammetric Engineering and Remote Sensing, 59*, 1491–1496.

Eastman, J.R., Van Fossen, M.E., and Solarzano, L.A. (2005). Transition potential modeling for land cover change. In D. Maguire, M. Goodchild and M. Batty (Eds.), *GIS, Spatial Analysis and Modeling*. ESRI Press, Redlands, CA. 413p.

Ehlers, M., Gahler, M., and Janowsky, R. (2003). Automated analysis of ultra high resolution remote sensing data for biotope type mapping: new possibilities and challenges. *ISPRS Journal of Photogrammetry and Remote Sensing, 57*, 315–326.

Ekstrand, S.P. (1990). Detection of moderate damage on Norway Spruce using Landsat TM and digital stand data. *IEEE Transactions on Geoscience and Remote Sensing, 28*, 685–692.

Emch, M., Quinn, J., Petersen, M., and Alexander, M. (2005). Forest cover change in the Toledo District, Belize from 1975 to 1999: a remote sensing approach. *The Professional Geographer, 57*, 256–267.

Florinsky, I.V. (1998). Combined analysis of digital terrain models and remotely sensed data in landscape investigations. *Progress in Physical Geography, 22*, 33–60.

Foody, G.M. (2002). Status of land cover classification accuracy assessment. *Remote Sensing of Environment, 80*, 185–201.

Foster, D., Motzkin, G., and Slater,B. (1998). Land-use history as long-term broad-scale disturbance: regional forest dynamics in central New England. *Ecosystems, 1*, 96–119.

Frank, T.D. (1988). Mapping dominant vegetation communities in the Colorado Rocky Mountain front range with Landsat Thematic Mapper and digital terrain data. *Photogrammetric Engineering and Remote Sensing, 54*, 1727–1734.

Franklin, J. (1995). Predictive vegetation mapping: geographic modeling of biospatial patterns in relation to environmental gradients. *Progress in Physical Geography, 19*, 474–499.

Franklin, J., Logan, T.L., Woodcock, C.E., and Strahler, A. (1986). Coniferous forest classification and inventory using Landsat and digital terrain data. *IEEE Transactions on Geoscience and Remote Sensing, 24*, 139–149.

Franklin, S.E., Moskal, L.M., Lavigne, M.B., and Pugh, K. 2000, Interpretation and classification of partially harvested forest stands in the Fundy model forest using multitemporal Landsat TM digital data. Canadian Journal of Remote Sensing 26(4):318–333.

Franklin, J., Phinn, S.R., Woodcock, C.E., and Rogan, J. (2003). Rationale and conceptual framework for classification approaches to assess forest resources and properties. In M.A. Wulder and S.E. Franklin (Eds.), *Methods and Applications for Remote Sensing of Forests: Concepts and Case Studies* (pp. 279–300). Kluwer Academic Publishers, Dordrecht, The Netherlands. 519p.

Franklin, J., Woodcock, C.E., and Warbington, R. (2000). Digital vegetation maps of forest lands in California integrating satellite imagery, GIS modeling, and field data in support of resource management. *Photogrammetric Engineering and Remote Sensing, 66*, 1209–1217.

Franklin, S.E. (2001). *Remote Sensing for Sustainable Forest Management.* Lewis Publishers, Boca Raton, FL. 407p.

Franklin, S.E. and Raske, A.G. (1994). Satellite remote sensing of spruce budworm forest defoliation in western Newfoundland. *Canadian Journal of Remote Sensing, 20*, 37–48.

Franklin, S.E. and Wulder, M.A. (2002). Remote sensing methods in medium spatial resolution satellite data land cover classification of large areas. *Progress in Physical Geography, 26*, 173–205.

Franklin, S.E., Wulder, M.A., Skakun, R., and Carroll, A. (2003). Mountain pine beetle red-attack forest damage classification using stratified Landsat TM data in British Columbia. *Photogrammetric Engineering and Remote Sensing, 69*, 283–288.

Friedl, M., Brodley, C., and Strahler, A. (1999). Maximizing land cover classification accuracies produced by decision trees at continental to global scales. *IEEE Transactions on Geoscience and Remote Sensing, 37*, 969–977.

Fuller, R.M., Smith, G.M., and Devereux, B.J. (2003). The characterization and measurement of land cover change through remote sensing: problems in operational applications? *International Journal of Applied Earth Observation and Geoinformation, 4*, 243–253.

Gahegan, M. (2003). Is inductive machine learning just another wild goose (or might it lay the golden egg)? *International Journal of Geographical Information Science, 17*, 69–92.

Gahegan, M. and Ehlers, M. (2000). A framework for the modelling of uncertainty between remote sensing and geographic information systems. *ISPRS Journal of Photogrammetry and Remote Sensing, 55,* 176–188.

Gahegan, M. and Flack, J. (1999). The integration of scene understanding within a geographic information system: a prototype approach for agricultural applications. *Transactions in GIS, 3,* 31–49.

Gemmell, F. and Varjo, J. (1999). Utility of reflectance model inversion versus two spectral indices for estimating biophysical characteristics in boreal forest test site. *Remote Sensing of Environment, 68,* 95–111.

Gemmell, F., Varjo, J., Strandstrom, M., and Kuusk, A. (2002). Comparison of measured boreal forest characteristics with estimates from TM data and limited ancillary information using reflectance model inversion. *Remote Sensing of Environment, 81,* 365–377.

Gong, P. and Xu, B. (2003). Remote sensing of forests over time: change types, methods, and opportunities. In M.A. Wulder and S.E. Franklin (Eds.), *Methods and Applications for Remote Sensing of Forests: Concepts and Case Studies* (pp. 301–333). Kluwer Academic Publishers, Dordrecht, The Netherlands. 519p.

Gopal, S., Woodcock, C.E., and Strahler, A. (1999). Fuzzy neural network classification of global land cover from a 1° AVHRR data set. *Remote Sensing of Environment, 67,* 230–243.

Guisan, A. and Zimmerman, N. (2000). Predictive habitat distribution models in ecology. *Ecological Modelling, 135,* 147–186.

Hansen, M.C., DeFries, R.S., Townshend, J.R.G., and Sohlberg, R. (2000). Global land cover classification at 1 km spatial resolution using a classification tree approach. *International Journal of Remote Sensing, 21,* 1331–1364.

Hansen, M.C., Dubayah, R., and DeFries, R.S. (1996). Classification trees: an alternative to traditional land cover classifiers. *International Journal of Remote Sensing, 17,* 1075–1081.

Hansen, M.C. and Reed, B.C. (2000). A comparison of the IGBP DISCover and University of Maryland Global Land Cover Products. *International Journal of Remote Sensing, 21,* 1365–1374.

Heikkonen, J. (2004). Forest change detection applying Landsat Thematic Mapper difference features: a comparison of different classifiers in Boreal forest conditions. *Forest Science, 50,* 579–588.

Herrmann, K., Rock, B.N., Ammer, U., and Paley, H.N. (1988). Preliminary assessment of airborne spectrometer and airborne thematic mapper data acquired for forest decline areas in the Federal Republic of Germany. *Remote Sensing of Environment, 24,* 129–149.

Hinton, J. (1999). Image classification and analysis using integrated GIS. In P. Atkinson and N. Tate (Eds.), *Advances in Remote Sensing and GIS Analysis* (pp.207–239). John Wiley and Sons, Chichester, U.K.

Holmes, K.W., Chadwick, O.A., and Kyriakidis, P.C. (2000). Error in a USGS 30-meter digital elevation model and its impact on terrain modeling. *Journal of Hydrology, 233,* 154–173.

Homer, C. and Gallant, A. (2001). *Partitioning the Conterminous United States into Mapping Zones for Landsat TM Land Cover Mapping.* USGS Draft White paper. Retrieved from http://landcover.usgs.gov. June 2004.

Huang, C., Homer, C., and Yang, L. (2003). Regional forest land-cover characterizations using medium spatial resolution satellite data. In M.A. Wulder and S.E. Franklin (Eds.), *Methods and Applications for Remote Sensing of Forests: Concepts and Case Studies* (pp. 389–410). Kluwer Academic Publishers, Dordrecht, The Netherlands. 519p.

Huang, X. and Jensen, J.R. (1997). A machine-learning approach to automated knowledge-base building for remote sensing image analysis with GIS data. *Photogrammetric Engineering and Remote Sensing, 63,* 1185–1194.

Huggard, D., Klenner, W., and Vyse, A., (1999). Windthrow following four harvest treatments in an engelmann spruce-subalpine fir forest in southern interior British Columbia, Canada. *Canadian Journal of Forest Research, 29*, 1547–1556.

Hunter, G. and Goodchild, M.F. (1997). Modeling the uncertainty of slope and aspect estimates derived from spatial databases. *Geographical Analysis, 29*, 35–49.

Hutchinson, C.F. (1982). Technique for combining Landsat and ancillary data for digital classification improvement. *Photogrammetric Engineering and Remote Sensing, 48*, 123–130.

Irland, L.C. (1998). Ice storm 1998 and the forests of the Northeast: A preliminary assessment. *Journal of Forestry, 96*, 32–40.

Jacobberger-Jellison, P.A. (1994). Detection of post-drought environmental conditions in the Tombouctou region. *International Journal of Remote Sensing, 15*, 3138–3197.

Jensen, J.R. (2005). *Digital Image Processing.* Prentice Hall, Upper Saddle River, NJ. 526p.

Johnson, R.D. (1994). Change vector analysis for disaster assessment: a case study of hurricane Andrew. *Geocarto International, 1*, 41–45.

Joria, P.E., Ahearn, S.C., and Connor, M. (1991). A comparison of the SPOT and Landsat Thematic Mapper satellite systems for detecting gypsy moth defoliation in Michigan. *Photogrammetric Engineering and Remote Sensing, 57*, 1605–1612.

Kasischke, E.S., Goetz, S., Hansen, M.C., Ozdogan, M., Rogan, J., Ustin, S., and Woodcock, C.E. (2004). Temperate and Boreal Forests. In S. Ustin (Ed.), *Manual of Remote Sensing. Vol. 4: Remote Sensing for Natural Resource Management and Environmental Monitoring.* John Wiley and Sons, New York. 848p.

Key, C.H. and Benson, N.C. (2002). Measuring and remote sensing of burn severity. In J.L. Coffelt and R.K. Livingston (Eds.), *U.S. Geological Survey Wildland Fire Workshop.* Los Alamos, NM, October 31–November 3, 2000. USGS Open-File Report 02-11. 55p. U.S. Department of The Interior, USGS, Denver, CO.

Khorram, S., Brockhaus, J., Bruck, R., and Campbell, M. (1990). Modeling and multitemporal evaluation of forest decline with Landsat TM digital data. *IEEE Transactions on Geoscience and Remote Sensing, 28*, 745–749.

Kittredge, D.B., Finley, A.O., and Foster, D.R. (2003). Timber harvesting as ongoing disturbance in a landscape of diverse ownership. *Forest Ecology and Management, 180*, 425–442.

Lamar, W.R., McGraw, J.B. and Warner, T.A. (2005). Multitemporal censusing of a population of eastern hemlock (*Tsuga canadensis* L.) from remotely sensed imagery using an automated segmentation and reconciliation procedure. *Remote Sensing of Environment, 95*, 133–143.

Lambin, E.F. (1999). Monitoring forest degradation in tropical regions by remote sensing: some methodological issues. *Global Ecology and Biogeography Letters, 8*, 191–198.

Lawrence, R.L., and Wright, A. (2001). Rule-based classification systems using classification and regression tree (CART) analysis. *Photogrammetric Engineering and Remote Sensing, 67*(10), 1137–1142.

Lees, B. (1996). Sampling strategies for machine learning using GIS. In M.F. Goodchild, L. Steyart, B. Parks, M. Crane, C. Johnston, D. Maidment, and S. Glendinning (Eds.), *GIS and Environmental Modelling: Progress and Research Issues* (pp. 39–42). GIS World, Ft. Collins, CO.

Lees, B.G. and Ritman, K. (1991). Decision-tree and rule-induction approach to integration of remotely sensed and GIS data in mapping vegetation in disturbed or hilly environments. *Environmental Management, 15*, 823–831.

Levien, L., Fischer, C., Roffers, P., Maurizi, B., Suero, J., Fischer, C., and Huang, X. (1999). *A Machine Learning Approach to Change Detection Using Multi-scale Imagery.* ASPRS Annual Conference, Portland, OR. (American Society for Photogrammetry and Remote Sensing, Bethesda, MD.)

Levien, L.M., Fischer, C., Roffers, P., Maurizi, B., and Suero, J. (2002). *Monitoring Land Cover Changes in California: Northeastern California Project Area*. State of California, Resources Agency, Department of Forestry and Fire Protection, Sacramento, CA, 171p. URL: http://frap.cdf.ca.gov/projects/land_cover/monitoring/pdfs/north eastern_ca_report.pdf (accessed May 26, 2006).

Lim, K., Treitz, P., Wulder, M., St-Onge, B., and Flood M. (2003). LiDAR remote sensing of forest structure. *Progress in Physical Geography, 27,* 88–106.

Lindemann, J.D. and Baker, W.L. (2002). Using GIS to analyze a severe forest blowdown in the southern Rocky Mountains. *International Journal of Geographic Information Science, 16,* 377–399.

Liu, J., Zhuang, D., Luo, D., and Xiao, X. (2003). Land-cover classification of China based on integrated analysis of AVHRR imagery and geo-spatial data. *International Journal of Remote Sensing, 24,* 2485–2500.

Loveland, T.R., Sohl, T.L., Stehman, S.V., Gallant, A.L., Sayler, K.L., and Napton, D.E. (2002). A strategy for estimating the rates of recent United States land-cover changes. *Photogrammetric Engineering and Remote Sensing, 68,* 1091–1100.

Luther, J.E., Franklin, S.E., Hudak, J., and Meades, J.P. (1997). Forecasting the susceptibility and vulnerability of balsam fir stands to insect defoliation with Landsat Thematic Mapper data. *Remote Sensing of Environment, 59,* 77–91.

Macomber, S.A. and Woodcock, C.E. (1994). Mapping and monitoring conifer mortality using remote sensing in the Lake Tahoe Basin. *Remote Sensing of Environment, 50,* 255–266.

Malerba, D., Esposito, F., Lanza, A., and Lisi, F.A. (2001). Machine learning for information extraction from topographic maps. In H.J. Miller and J. Han (Eds.), *Geographic Data Mining and Knowledge Discovery* (pp.291–314). Taylor and Francis, New York.

McCullagh, P. and Nelder, J. (1989). *Generalized Linear Models*. Chapman and Hall, London, U.K. 511p.

McIver, D.K. and Friedl, M.A. (2002). Using prior probabilities in decision-tree classification of remotely sensed data. *Remote Sensing of Environment, 81,* 253–261.

McIver, D. and Wheaton, E., (2005). Tomorrow's forests: adapting to a changing climate. *Climatic Change, 70,* 273–282.

Medler, M. and Yool, S.R. (1997). Improving Thematic Mapper based classification of wildfire induced vegetation mortality. *Geocarto International, 12,* 49–58.

Mertens, B., Forni, E., and Lambin, E.F. (2001). Prediction of the impact of logging activities on forestcover: a case-study in the East province of Cameroon. *Journal of Environmental Management, 62,* 21–36.

Millward, A.A., and Kraft, C.E. (2004). Physical influences of landscape on a large-extent ecological disturbance: the northeastern North American ice storm of 1998. *Landscape Ecology, 19*(1): 99–111.

Mladenoff, D.J. (2005). The promise of landscape modeling: successes, failures, and evolution. In J.A. Wiens, L. Fahrig, B. Milne, P. Dennis, R. Hobbs, J. Nassauer, and M. Moss (Eds.), *Issues and Perspectives in Landscape Ecology*. Cambridge University Press, Cambridge, U.K. 390p.

Moore, I., Grayson, R., and Ladson, A. (1991). Digital terrain modelling: a review of hydrological, geomorphological, and biological applications. *Hydrological Processes, 5,* 3–30.

Morisette, J.T., Khorram, S., and Mace, T. (1999). Land-cover change detection enhanced with generalized linear models. *International Journal of Remote Sensing, 20,* 2703–2721.

Muchoney, D. and Williamson, J. (2001). A Gaussian adaptive resonance theory neural network classification algorithm applied to supervised land cover mapping using multitemporal vegetation index data. *IEEE Transactions on Geoscience and Remote Sensing, 39,* 1969–1977.

National Center for Geographic Information and Analysis. (2005). Retrieved from: http://www.ncgia.ucsb.edu. Website: Retrieved June 2004.

Nelson, R.F. (1983). Detecting forest canopy change due to insect activity using Landsat MSS. *Photogrammetric Engineering and Remote Sensing, 49*, 1303–1314.

Nepstad, D.A., Veríssimo, A., Alencar, A., Nobre, C., and Lima, E. (1999). Large-scale impoverishment of Amazonian forests by logging and fire. *Nature, 398*, 505–508.

Nilson, T., Olsson, H., Anniste, J., Lükk, T., and Praks, J. (2001). Thinning-caused change in reflectance of ground vegetation in boreal forest. *International Journal of Remote Sensing, 22*, 2763–2776.

Olsson, H. (1995). Reflectance calibration of Thematic Mapper data for forest change detection. *International Journal of Remote Sensing, 50*, 221–230.

Olthof, I., King, D., and Lautenschlager, R.A. (2004). Mapping deciduous storm damage using Landsat and environmental data. *Remote Sensing of Environment, 89*, 484–496.

Pan, D., Domon, D., Marceau, D., and Bouchard, A. (2001). Spatial pattern of coniferous and deciduous forest patches in an eastern North America agricultural landscape: the influence of land use and physical attributes. *Landscape Ecology, 16*, 99–110.

Parmenter, A.W., Hansen, A., Kennedy, R.E., Cohen, W., Langer, U., Lawrence, R., Maxwell, B., Gallant, A., and Aspinall, R. (2003). Land use and land cover change in the greater Yellowstone ecosystem. *Ecological Applications, 13*(3), 687–703.

Pellikka, P., Seed, E.D., and King, D.J. (2000). Modelling deciduous forest ice storm damage using aerial CIR imagery and hemispheric photography. *Canadian Journal of Remote Sensing, 36*, 394–405.

Perlitsch, S. (1995). Terrain Shape Classification of Digital Elevation Models Using Eigenvectors and Fourier Transforms. Ph.D. dissertation, SUNY-Syracuse, Syracuse, NY.

Peters, A.J., Reed, B.C., Eve, M.D., and Havstad, K.M. (1993). Satellite assessment of drought impact on native plant communities of southeastern New Mexico, U.S.A. *Journal of Arid Environments, 24*, 305–319.

Peterson, S.H., and Stow, D.A. (2003). Using multiple image endmember spectral mixture analysis to study chaparral regrowth in southern California. *International Journal of Remote Sensing, 24*(22), 4481–4504.

Pontius, R.G., Jr. (2002). Statistical methods to partition effects of quantity and location during comparison of categorical maps at multiples resolutions. *Photogrammetric Engineering and Remote sensing, 68*, 1041–1049.

Powell, R.L., Matzke, N., de Souza, C., Jr., Clark, M., Numata, I., Hess, L.L., and Roberts, D.A. (2004). Sources of error in accuracy assessment of thematic land-cover maps in the Brazilian Amazon. *Remote Sensing of Environment, 90*, 221–234.

Pyne, S.J., Andrews, P.L., and Laven, R.D. (1996). *Introduction to Wildland Fire*. John Wiley and Sons, New York. 808p.

Quinlan, J.R. (1993). *C4.5: Programs for Machine Learning*. Morgan Kaufman, San Mateo, CA.

Raclot, D., Colin, F., and Puech, C. (2005). Updating land cover classification using a rule-based decision system. *International Journal of Remote Sensing, 26*, 1309–1321.

Radeloff, V.C., Mladenoff, D.J., and Boyce, M.S. (1999). Detecting jack pine budworm defoliation using spectral mixture analysis: separating effects from determinants. *Remote Sensing of Environment, 69*, 156–169.

Ramsey, R.D., Falconer, A., and Jensen, J.R. (1995). The relationship between NOAA-AVHRR NDVI and ecoregions in Utah. *Remote Sensing of Environment, 53*, 188–198.

Raza, A. and Kaiz, R. (2001). An object-oriented approach for modeling urban land-use changes. *URISA Journal, 14*, 37–55.

Riano, D., Chuvieco, E., Ustin, S., Zomer, R., Dennison, P., Roberts, D., and Salas, J. (2002). Assessment of vegetation regeneration after fire through multitemporal analysis of AVIRIS images in the Santa Monica Mountains. *Remote Sensing of Environment,* 79:60–71.

Ricchetti, E. (2000). Multispectral satellite image and ancillary data integration for geological classification. *Photogrammetric Engineering and Remote Sensing, 66,* 429–436.

Rogan, J. and Chen, D. (2004). Remote sensing technology for mapping and monitoring land cover and land use change. *Progress in Planning, 61,* 301–325.

Rogan, J. and Franklin, J. (2001). Mapping wildfire burn severity in southern California forests and shrublands using enhanced Thematic Mapper imagery. *Geocarto International, 16,* 89–99.

Rogan, J., Miller, J., Stow, D., Franklin, J., Levien, L., and Fischer, C. (2003). Land cover change mapping in California using classification trees with Landsat TM and ancillary data. *Photogrammetric Engineering and Remote Sensing, 69,* 793–804.

Rogan, J. and Yool, S.R. (2001). Mapping fire-induced vegetation depletion in the Peloncillo Mountains, Arizona and New Mexico. *International Journal of Remote Sensing, 22,* 3101–3121.

Rohde, W.G. and Moore, H.J. (1974). Forest defoliation assessment with satellite imagery. *Proceedings of the Ninth International Symposium on Remote Sensing of Environment,* Ann Arbor, MI, April. 15–19. Environmental Research Institute of Michigan, Ann Arbor, MI.

Roy, D.P., Giglio, L., Kendall, J.D., and Justice, C.O. (1999). A multitemporal active-fire based burn scar detection algorithm. *International Journal of Remote Sensing, 20,* 1031–1038.

Sader, S.A., Bertrand, M., and Wilson, E.H. (2003). Satellite change detection patterns on an industrial forest landscape. *Forest Science, 49,* 341–353.

Saksa, T., Uuttera, J., Kolström, T., Lehikoinen, M., Pekkarinen, A., and Sarvi, V. (2003). Clearcut detection in boreal forest aided by remote sensing. *Scandinavian Journal of Forest Research, 18,* 537–546.

Satterwhite, M., Rice, W., and Shipman, J. (1984). Using landform and vegetation factors to improve the interpretation of LANDSAT imagery. *Photogrammetric Engineering and Remote Sensing, 50,* 83–91.

Saveliex, A.A. and Dobrinin, D.V. (2002). The use of Kohonen's neural nets for the detection of land-cover transitions. In L. Bruzzone and P. Smits (Eds.), *First Annual Workshop on the Analysis of Multi-temporal Remote Sensing Images.* World Scientific Publishing Co. 440p. Hackensack, NJ: (p. 148–155)

Schwarz, M. and Zimmerman, N. (2005). GLM-based forest continuous fields- an optimization based on MODIS data for complex topographic terrain. *Remote Sensing of Environment, 95,* 428–443.

Senoo, T., Kobayashi, F., Tanaka, S., and Sugimura, T. (1990). Improvement of forest type classification by SPOT HRV with 20 m mesh DTM. *International Journal of Remote Sensing, 11,* 1011–1022.

Seto, K., Woodcock, C.E., Song, C., Huang, X., Lu, J., and Kaufmann, R. (2002). Monitoring land-use change in the Pearl River Delta using Landsat TM. *International Journal of Remote Sensing, 21,* 1985–2004.

Shasby, M. and Carneggie, D. (1986). Vegetation and terrain mapping in Alaska using Landsat MSS and digital terrain data. *Photogrammetric Engineering and Remote Sensing, 52,* 779–787.

Siqueira, P., Chapman, B., Saatchji, S., and Freeman, A., 1997, Amazon Rainforest visualization/classification enabled by supercomputers (ARVORES). Proceedings of the International Geoscience and Remote Sensing Symposium (IGARSS'97), Singapore, 3–8 August, 1997, volume 1, pp. 104–106, IEEE, Piscataway NJ.

Siqueira, P., Chapman, B., and McGarragh, G. (2003). The coregistration, calibration and interpretation of the multiseason JERS-1 SAR data over South America. *Remote Sensing of Environment, 87*, 389–403.

Siqueira, P., Hensley, S. Shaffer, S., Hess, L., McGarragh, G., Chapman, B. and Freeman, A. (2000). A continental scale mosaic of the Amazon Basin using JERS-1 SAR. *IEEE Transactions on Geosciences and Remote Sensing, 38*, 2638–2644.

Skakun, R., Franklin, S.E., and Wulder, M.A. (2003). Sensitivity of the EWDI to mountain pine beetle red-attack damage. *Remote Sensing of Environment, 30*, 433–443.

Skidmore, A. (1989). A comparison of techniques for calculating gradient and aspect from a gridded digital elevation model. *International Journal of Geographical Information Science, 3*, 323–334.

Skidmore, A.K. and Turner, B.J. (1988). Forest mapping accuracies are improved using a supervised nonparametric classifier with SPOT data. *Photogrammetric Engineering and Remote Sensing, 54*, 1415–1421.

Smith, A.M.S., Wooster, M.J., Drake, N.A.,. Dipotso, F.M., Falkowski, M.J., and Hudak, A.T. (2005). Testing the potential of multi-spectral remote sensing for retrospectively estimating fire severity in African Savannahs. *Remote Sensing of Environment, 97*, 92–115.

Smith, J., Lin, T., and Ranson, K. (1980). The Lambertian assumption and Landsat data. *Photogrammetric Engineering and Remote Sensing, 46*, 1183–1189.

Smith, J.L., Clutter, M., Keefer, B., and Ma, Z. (2003). The future of digital remote sensing for production forestry organizations. *Forest Science, 49(3)*, 455–456.

Song, M. and Civco, D. (2004). Road extraction using SVM and image segmentation. *Photogrammetric Engineering and Remote Sensing, 70*, 1365–1372.

Souza, C.M. and Barreto, P. (2000). An alternative approach for detecting and monitoring selectively logged forests in the Amazon. *International Journal of Remote Sensing, 21*, 173–179.

Souza, C.M., Roberts, D.A., and Cochrane, M.A. (2005). Combining spectral and spatial information to map canopy damage from selective logging and forest fires. *Remote Sensing of Environment, 98*, 329–343.

Steyaert, L.T. (1996). Status of land data for environmental modeling and challenges for geographic information systems in land characterization. In M.F. Goodchild, L. Steyart, B. Parks, M. Crane, C. Johnston, D. Maidment, and S. Glendinning (Eds.), *GIS and Environmental Modelling: Progress and Research Issues* (pp. 17–27). GIS World, Ft. Collins, CO. 504p.

Stoms, D. (1996). Validating large-area land cover databases with maplets. *Geocarto International, 11*, 87–95.

Stow, D. (1999). Reducing misregistration effects for pixel-level analysis of landcover change. *International Journal of Remote Sensing, 20*, 2477–2483.

Strahler, A. (1980). The use of prior probabilities in maximum likelihood classification of remote sensing data. *Remote Sensing of Environment, 10*, 135–163.

Strahler, A., Logan, T.L., and Bryant, N.A. (1978). Improving forest classification accuracy from Landsat by incorporating topographic data. *Twelfth International Symposium on Remote Sensing of Environment* (pp. 927–942). Ann Arbor, MI.

Talbot, S.S. and Markon, C.J. (1986). Vegetation mapping of Nowitna National Wildlife Refuge, Alaska using Landsat MSS Digital Data. *Photogrammetric Engineering and Remote Sensing, 52*, 791–799.

Trietz, P. and Howarth, P. (2000). Integrating spectral, spatial and terrain variables for forest ecosystem classification. *Photogrammetric Engineering and Remote Sensing, 66(3)*, 305–317.

Treitz, P. and Rogan, J. (2004). Remote sensing for mapping and monitoring land cover and land use change: an introduction. *Progress in Planning, 61*, 269–279.

Turner, B.L., II, Skole, D., Sanderson, S., Fischer, G., Fresco, L.O., and Leemans, R. (1999). *Land-Use and Land-Cover Change Science/Research Plan.* International Geosphere-Biosphere Programme, Stockholm, Sweden. IGBP Report No. 35 and HDP Report No. 7.

Turner, B.L., II. (2002). Toward integrated land-change science: advances in 1.5 decades of sustained international research on land-use and land-cover change. In W. Steffen, J. Jäger, D. Carson, and C. Bradshaw (Eds.), *Challenges of a Changing Earth: Proceedings of the Global Change Open Science Conference* (pp. 21–26). Amsterdam, The Netherlands, July 10–13, 2001. Springer-Verlag, Heidelberg, Germany.

Ustin, S., Zarco-Tejada, P.J., Jacquemoud, S., and Asner, G.P. (2004). Remote sensing of the environment: state of the science and new directions. In S. Ustin (Ed.), *Manual of Remote Sensing Volume 4: Remote Sensing for Natural Resource Management and Environmental Monitoring.* John Wiley and Sons, New York. 848p.

Van Niel, K., Laffan, S., and Lees, B. (2004). Effect of error in the DEM on environmental variables for predictive vegetation modelling. *Journal of Vegetation Science, 15*, 747–756.

Varjo, J. (1997). Change detection and controlling forest information using multitemporal Landsat TM imagery. *Acta Forestalia Fennica, 258*, 1–64.

Vigilante, T., Bowman, D.M.J.S., Fisher, R., Russell-Smith, J., and Yates, C., (2004). Contemporary landscape burning patterns in the far North Kimberly region of north-west Australia: human influences and environmental determinants. *Journal of Biogeography, 31*, 1317–1333.

Vogelmann, J.E., Sohl, T.L., Campbell, P.V., and Shaw, D.M. (1998). Regional land cover characterization using LANDSAT Thematic Mapper data and ancillary data sources. *Environmental Monitoring and Assessment, 51*, 415–428.

Warner, T.A., Levandowski, D.W., Bell, R., and Cetin, H. (1994). Rule-based geobotanical classification of topographic, aeromagnetic and remotely sensed vegetation community data. *Remote Sensing of Environment, 50*, 41–51.

White, J., Ryan, K., Key, C., and Running, S. (1996). Remote Sensing of forest fire severity and vegetation recovery. *International Journal of Wildland Fire, 6*, 125–136.

White, J.C., Wulder, M.A., Brooks, D., Reich, R., and Wheate, R.D. (2005). Detection of red attack stage mountain pine beetle infestation with high spatial resolution satellite imagery. *Remote Sensing of Environment, 96*, 340–351.

Williams, D.L. and Nelson, R.F. (1986). Use of remotely sensed data for assessing forest stand conditions in the Eastern United States. *IEEE Transactions on Geoscience and Remote Sensing, 24*, 130–138.

Wilson, E.H. and Sader, S.A. (2002). Detection of forest harvest type using multiple dates of Landsat TM imagery. *Remote Sensing of Environment, 80*, 385–396.

Wilson, J.P. and Gallant, J.C. (1998). Terrain-based approaches to environmental resource evaluation. In S.N. Lane, K.S. Richards, and J.H. Chandler (Eds.), *Landform Monitoring, Modelling and Analysis* (pp. 219–240). John Wiley and Sons, Chichester, U.K.

Woodcock, C.E. (2002). Uncertainty in remote sensing. In G. Foody and P. Foody (Eds.), *Uncertainty in Remote Sensing and GIS* (pp. 19–24). John Wiley and Sons, New York.

Wulder, M. (1998). Optical remote sensing techniques for the assessment of forest inventory and biophysical parameters. *Progress in Physical Geography, 22*, 449–476.

Wulder, M.A., Franklin, S.E., White, J., Cranny, M., and Dechka, J. (2004). Inclusion of topographic attributes in an unsupervised classification of satellite imagery. *Canadian Journal of Remote Sensing, 30*, 137–149.

Wulder, M.A., Skakun, R., Franklin, S.E., and White, J. (2005). Enhancing forest inventories with mountain pine beetle infestation information. *Forestry Chronicle, 81*, 149–159.

Yool, S.R., Eckhardt, D., Estes, J., and Cosentino, M. (1985). Describing the brushfire hazard in Southern California. *Annals of the Association of American Geographers, 75*, 417–430.

Zhan, X., Sohlberg, R., Townshend, J.R.G., DiMiceli, C., Carroll, M., Eastman, J.C., Hansen, M.C., and DeFries, R.S. (2002). Detection of land cover changes using MODIS 250 meter data. *Remote Sensing of Environment, 83*, 336–350.

7 New Directions in Landscape Pattern Analysis and Linkages with Remote Sensing

Sarah E. Gergel

CONTENTS

INTRODUCTION

Spatial patterning in forests and landscapes has long been of interest to ecologists, foresters, and managers. From the "natural" forces that shape landscapes (such as fire and insect outbreaks) to the cultural and anthropogenic forces that shape landscapes (road building, urbanization, and harvesting), the quantification of these patterns has been a major focus of much of ecological and management-related research. Many of the reasons for such interest in the quantification of spatial pattern have been elaborated on in previous chapters in this volume. The technological and conceptual advances in remote sensing have shaped the way landscape ecologists (the inventors of pattern analysis techniques discussed here) conduct research. It is likely that the discipline of remote sensing will continue to wield this important influence. In this chapter, the interplay between remote sensing and pattern analysis (using landscape metrics or landscape pattern indices) is emphasized in an attempt not only to provide context for how pattern analysis is currently conducted but also to explore the ways in which pattern analysis might develop in the future.

The study of spatial pattern has progressed rapidly in only a matter of decades, from the early days of spatial analysis (the 1970s) when paper maps were simply overlaid with overheads and plasticine layers of other maps, to the automated routines of geographical information system (GIS) and image analysis software that can generate hundreds of landscape pattern indices (LPIs) in minutes. The discipline of remote sensing and related advances and techniques to a large extent can be credited for the ability to ask questions about spatial pattern at broad scales. The ability to ask questions at the broader scales on large landscapes (tens of thousands of kilometers) as opposed to the fine, plot-level scale of traditional ecology and routine field sampling was a major advance in spawning the discipline of landscape ecology. The fact that such databases as Landsat are familiar to, and routinely used by, those in a wide variety of environmental sciences is a major contribution to the environmental sciences. However, as pattern analysis software and imagery become more widely available, the potential for misuse of these tools and data are nontrivial.

Interestingly, the latest advances in remote sensing techniques not only allow the analysis of these broad regions of interest to landscape ecology, but also are now producing fine-scale, submeter resolution data sources (e.g., using IKONOS, Quick-Bird). This fine scale was until recently the purview of field-sampling programs and the mainstay of traditional field ecology, forestry professionals, and field managers (Wulder, Hall, et al., 2004). Because of this new convergence of spatial scales, it is useful to assess the philosophy and lessons learned from spatial pattern analysis within the context of the future directions of remote sensing that will include these finer spatial resolution sources of data. While aerial photos and GIS also largely helped shape ecological research and management, spatial pattern analysis was largely driven by a "Landsat view" of the world, given its ubiquitous and extensive coverage and compatibility with raster-based LPI techniques.

The goals of this chapter are to examine the linkages between remote sensing and spatial pattern analysis and how they might both evolve to assist in the assessment of landscape change. First, the basic concepts of pattern analysis are discussed. This discussion also examines pattern analysis within the context of our prevailing conceptual notions of fragmentation, which have advanced considerably over time, as well as discusses issues of relevance to quantifying natural levels of variability in fragmentation and disturbance in forested ecosystems. While maintaining a focus on raster data, several questions are posed that could benefit pattern analysis greatly through more collaborative work between remote sensing scientists and landscape ecologists.

HOW DO WE MEASURE SPATIAL PATTERN?

Before delving into the implications of new remote sensing techniques for pattern analysis, a review of some of the major concepts in pattern analysis is warranted. Readers familiar with the tools of pattern analysis from the perspective of landscape ecology might consider skipping to the next section. Those interested in "hands-on" experience with pattern analysis software (using pre-prepared data sets and free software) may benefit from examining the work of Cardille and Turner (2001) and Greenberg et al. (2001).

IMPORTANCE OF GRAIN, MINIMUM MAPPING UNIT, EXTENT, AND CLASSIFICATION SCHEME

For those relatively new to pattern analysis using landscape metrics or landscape pattern indices (Table 7.1), the primary goal is often to use a classified map and ask such basic questions as:

How many cover types are on this map?
What proportion of the map is occupied by old forest?
Which forest types are fragmented and dispersed?
Which cover types are highly connected?
How many remnant forest patches are there?
How many burned forest patches are there?

Interestingly (or unfortunately), the answers to the above questions are very much influenced by the grain, minimum mapping unit (MMU), extent, and classification scheme used to create the map. Furthermore, not only can quantitatively different answers be obtained, but also *qualitatively* different answers may be obtained using the exact same map with different grain, extent, and classification scheme. Similarly, changes through time in one location (i.e., the results from an analysis of change detection) may be erroneous if the maps that are compared differ in any of these aspects. This is a nontrivial point that is important to consider not only in one's own analysis but also in interpreting the results of others. Examples of the impact of these decisions are numerous, and a few generalities are relayed here.

Grain

Grain size is one aspect of scale and in the context of raster maps is typically analogous to the pixel size on a regularly partitioned square tessellation (or the

TABLE 7.1
Summary of Commonly Used Landscape Metrics and More Recently Developed Ones Discussed in This Chapter

Metric	Equation	Indicates	Reference
Proportion p_i	$$p_i = \frac{\sum_{j=1}^{n} a_{ij}}{A}(100)$$ where a_{ij} = area (m²) of patch ij; and A = total landscape area (m²)	Proportion of landscape occupied by class i; also termed PLAND in *Fragstats* s	Turner et al. (2001)
Total core area (TCA)	$$TCA = \sum_{j=1}^{n} a_{ij}^c \left(\frac{1}{10,000}\right)$$ a_{ij}^c = core area (m²) of patch ij based on specified edge depths (m)	The sum of the core areas of each patch of the corresponding patch type (divided by 10,000 to convert to hectares)	McGarigal et al. (2002)
Number of patches (NP)	NP = total number of patches in the landscape	Higher values indicate more fragmentation; also termed total clusters	McGarigal et al. (2002)
Mean patch size (MPS)	$$MPS = \frac{\sum_{i=1}^{i=NP} a^i}{NP}$$	Low values indicate high levels of fragmentation	McGarigal et al. (2002)
Area-weighted mean patch size (AWMPS)	$$AWMPS = \sum \frac{S_i^2}{S}$$ where S_i is the size of the largest patch	Larger values indicates less fragmentation; also termed S_{av} size	Gardner (2001)
Total edge (TE), edge length (EL), inner edge length	Variety of methods possible (areal and linear)	Higher values indicate more edge habitat, more fragmentation	Gardner (2001); McGarigal et al. (2002)

Metric	Equation	Behavior	Reference
Largest patch index (LPI)	$LPI = \dfrac{\max\limits_{j=1}^{n}(a_{ij})}{A}(100)$	LPI is the percentage of the landscape comprised by the largest patch; larger values indicate less fragmentation	McGarigal et al. (2002)
Mean shape index (MSI)	$MSI = \dfrac{\sum\limits_{i=1}^{i=NP} \dfrac{p_i}{4\sqrt{a^i}}}{NP}$	Higher values indicate greater complexity of shapes	Saura (2002)
Landscape division (LD)	$LD = 1 - \sum\limits_{i=1}^{i=NP}\left[\dfrac{a^i}{A^T}\right]^2$ where A^T is total area of landscape	Higher values indicate more fragmentation	Jaeger (2000)
Effective mesh size (EMS)	$EMS = A^T(1 - LD)$ Note: $EMS = \dfrac{A^1}{100}\,AWMPS$		Jaeger (2000)
Patch cohesion (PC)	$PC = \left[1 - \dfrac{\sum^p}{\sum(p\sqrt{a})}\right]\left[1 - \dfrac{1}{\sqrt{N}}\right]^{-1}$	Higher values are correlated with higher simulated dispersal success	Schumaker (1996)
Matheron fragmentation index	$M = \dfrac{\#\ \text{edges}}{\sqrt{\#\ \text{forest pixels}}\,\sqrt{\#\ \text{total pixels}}}$	Higher values indicate more fragmentation	Matheron (1970), with equation from Jeanjean and Achard (1997)

Calculations may vary depending on the software used and input specifications. This table is by no means exhaustive; a more comprehensive summary of metrics is available in the work of McGarigal et al. (2002), and the behavior is summarized in the work of Neel et al. (2004).

spatial resolution of the image data source). Earlier studies suggested that changes in landscape pattern indices due to scale changes may not be extremely problematic. For example, aerial photography used to create raster maps of varying pixels size (4-, 12-, 28-, to 80-m cell sizes) suggested that in some situations the effects of changing scale were not dramatic (Wickham and Riitters, 1995). Another early study found that while pattern indices were sensitive to change in grain, estimating land-scape pattern indices at different scales was fairly feasible using aggregation algo-rithms (Benson and MacKenzie, 1995).

Recent work suggests more serious complications than initially identified, with increases in grain size generally leading to the loss of the most rare or fragmented classes on a map. Interestingly, it is precisely such cover classes that are often of concern to conservation (rare habitat) or management (small patchy disturbances). This loss of the rarest classes is then accompanied by an increase in the most dominant cover classes. With increasing grain, the complexity of patches will likely decrease as the edges between classes will be underestimated (Turner et al., 2001).

A comparison of SPOT (Systeme Pour l'Observation de la Terre) multispectral high-resolution visible, Landsat Thematic Mapper (TM) and National Oceanic and Atmospheric Administration Advanced Very High Resolution Radiometer (AVHRR) suggested that indices such as landscape division and the largest patch index may be less sensitive to this type of scale change (Saura, 2004), whereas the number of patches, edge length, and mean patch size were quite sensitive to varying resolution and should *not* be compared among maps with varying resolution (Saura, 2004). Another fragmentation metric, patch cohesion, while sensitive to resolution at low proportions of remaining habitat, is less sensitive at higher proportions. Incidentally, because the metric is less sensitive to changes in spatial pattern (aggregation) at high proportions, it is less useful for quantifying habitat fragmentation (Saura, 2004). Thus, different metrics might respond quite markedly to changes in grain.

Due to the diversity of responses of metrics to changes in grain, studies have attempted to create scaling laws for translation of the results of a pattern analysis across scales, using both "real" and "simulated" landscapes (Shen et al., 2004; Wu, 2004). Shen et al. (2004) found several general groupings of pattern indices accord-ing to their scaling behavior and suggest the following groupings:

- Type I showed predictable behavior with simple scaling functions such as power laws or linear functions
- Type II exhibited stair-step response patterns and were essentially not predictable
- Type III showed unpredictable and erratic responses to scale changes

Other work has even attempted to use less-spatial, coarser-scale forest inventory (stand attribute data) data to predict the values of pattern indices at finer scales (Cumming and Vernier, 2002; Kleinn, 2000). Despite this work, reliable, well-tested, universal scaling laws, which are transportable among different regions, currently do not exist.

As noted by Saura (2004), most of the studies that examine the impact of changing grain cell size do so by "coarsening up" of an existing image (with exceptions, including Benson and MacKenzie, 1995 and Saura, 2004). Such artificial coarsening is important as scaling laws using artificially coarsened data (in contrast to spatial data from a coarser spatial resolution sensor) may be subject to the influence of significant aggregation errors due to the assumptions used for grouping pixels. For example, a majority rule filter may result in a more fragmented pattern (on the coarser version) than the pattern on an image derived from a coarser scale sensor (Saura, 2004). Differences in fragmentation metrics from maps created through aggregation (via majority rules) versus metrics from coarser-grained sensors suggest that the values for number of patches, mean patch size, and edge length may suggest a higher level of fragmentation in a landscape with cells that have been aggregated (Saura, 2004). The loss of information and errors resulting from the aggregation of fine-scale data into coarser scales have been well examined, with a useful overview provided by Turner et al. (2001). A variety of equations which are helpful in reducing the errors in the resulting areal estimates of cover types exist and there are rules that can be followed to ensure that the value of proportion is maintained even over several aggregations (Turner et al., 2001).

Nonetheless, even when comparing images at different scales that have not been artificially coarsened but are derived from different sensors, some important issues must be considered. An early study compared SPOT Landsat TM and AVHRR imagery classified into two classes using the near-infrared band to distinguish land and water (Benson and MacKenzie, 1995), with a high resultant accuracy (92% accuracy for water and 99% for land). As the pixel size changed from 20 m to 1.1 km, percentage water decreased by 44% (Benson and MacKenzie, 1995). Thus, the areal estimates obtained for cover classes obtained from different resolution sensors may also vary and must be accounted for when examining changes in pattern indices.

Minimum Mapping Unit

A related issue often confused with grain cell size is that of MMU. This is often referred to as the size of the smallest area that will be mapped as a distinct area. Often, a MMU is implemented in the postclassification phase as small remnant patches (below some minimum size) are removed through use of a majority (modal) filter or other technique. The rationale is to remove the speckled "salt-and-pepper" apparent in an image. In practice, this idea is often confused with grain (by nonremote sensing scientists) due to a failure to recognize that the MMU is used "post hoc" in a classification scheme. Maps with the same grain and extent but differing MMUs can have quite different resulting landscape metrics (Langford et al., 2006; Saura, 2002).

Rare and fragmented cover types may be lost when using an increased MMU, along with a concomitant increase in the most dominant cover classes (Saura, 2002). The implications of this procedure are directly relevant for any study that examines patch sizes because the smallest patches are assumed to be errors in this context and thus are removed from the image. Thus, simple measures of fragmentation such as the total number of patches and mean patch size should always be viewed with caution, with some explicit discussion of the grain and MMU implemented on the map from which the pattern indices are drawn.

Fragmentation metrics such as landscape division and related indices have been suggested for use when comparing fragmentation among images with different MMUs (Saura, 2002). It has been suggested that mean shape index in particular should not be used when comparing maps with different MMUs (or even a different patch size distribution) as it is most sensitive to MMU, while an area-weighted mean shape index is robust. However, the ideal situation would use images with identical MMUs. The frequency with which MMUs are used in the postprocessing of spatial data, and the fact that MMUs may differ among cover classes, combined with the frequency of use of patch level statistics for pattern analysis suggest that this issue should always be explicitly mentioned in any pattern analysis.

Extent

The spatial extent of an image, or the total area of an image, can have an impact on the results of a pattern analysis. An increase in the spatial extent examined often leads to an increase in the number of cover types observed (Turner et al., 2001). Smaller extents can lead to the truncation of the largest patch as well as changes in the shape of many other patches, so metrics of patch shape need to be used with caution in smaller maps. Furthermore, truncation of the map can lead to consistent, if not great, overestimation of the number of habitat patches as connections among adjacent patches near the edge of the map are severed (envision a large irregularly shaped patch near the edge of a landscape). One key heuristic is to avoid undertaking a pattern analysis on maps of less than 100 by 100 cells. Another heuristic is that the map extent should be two to five times greater than the size of the largest patch on the landscape (O'Neill et al., 1996). Extent does not necessarily co-vary with grain, but often the use of an extremely broad extent also results in the use of coarser grain data, so in practice, issues of extent and grain must often be considered simultaneously.

Not only is map extent important in terms of its absolute area, but also the logic or rationale used in demarcating landscape boundaries is quite important. In one multiscale analysis of land cover pattern, the use of arbitrary rectangular boundaries was compared to the results from landscapes demarcated using watershed boundaries (Cain et al., 1997). According to the pattern indices, landscapes defined by watershed boundaries were more homogeneous within, than landscapes defined by arbitrary rectangles (Cain et al., 1997). This was logically surmised to be related to the geophysical controls exerted over the region. Quantifying natural levels of variability in ecosystems (Landres et al., 1999) remains an important challenge for applied and basic research. Given the clear importance of the watershed as a management unit, characterizing heterogeneity and quantifying landscape patterns over the scale of watersheds should be encouraged further as an extent more appropriate than randomly demarcated landscapes that are estranged from the geophysical template on which the landscape developed. While there will rarely be one prescriptive "right" scale over which to conduct a pattern analysis, depending on the questions under consideration, certain scales (grain, MMU, and extents) are likely more appropriate than others. As such, careful consideration should always be given to these choices and assumptions.

Classification Scheme

It is important for any pattern analysis to discuss the origins and rationale used to create the classification used before that image is used in any pattern analysis. From the standpoint of end users of remotely sensed imagery, the classification scheme is among the most important variables to consider to interpret remotely sensed data and the resulting pattern analysis appropriately. Many end users have little control over this aspect as it is "predetermined" from the standpoint of most end users who are not remote sensing scientists. The preprocessing of the data into some predefined classification scheme affects every metric of pattern that is derived from an image. The rationale behind the classification scheme is among the important variables most ignored and not addressed directly in the translation of imagery to the final end user. Ultimately, problems arise when classification schemes developed for one application are used carte blanche by others for a different application.

It is not uncommon for simpler classification schemes (i.e., fewer categories) to have a higher classification accuracy. Extreme examples might include classifications depicting only land and water with accuracies exceeding 95% (Benson and MacKenzie, 1995), or a habitat versus nonhabitat classification scheme. Because the proportions of different cover classes can have an impact on all derived metrics, it is quite conceivable that the impact of classification errors on class proportion and thus many pattern indices could be reduced through the use of simpler classification schemes. An important caveat remains that a simpler classification scheme, if created through the use of a majority rules filter, could induce aggregation errors.

SPATIALLY IMPLICIT VERSUS SPATIALLY EXPLICIT MEASURES OF PATTERN

Spatially explicit measures of pattern yield information about the location of a class or patch in the landscape. In contrast, spatially implicit, or nonspatial, measures include such metrics as "percentage cover" or "total area" of a land cover class or p_i (the proportion of the landscape in class i). Such measures give no indication of where in the landscape such cover classes occur; the importance of this distinction cannot be overstated. Spatially implicit measures are straightforward to measure and are often used in the calculation of other metrics. Ecological or management-related questions that can rely on spatially implicit measures of land cover are simpler than those that require spatially explicit landscape metrics of configuration (Gergel, 2005). Also, nonspatial measures are the easiest to rectify with the most basic types of classification accuracy assessments used for remotely sensed imagery.

Landscape pattern indices of configuration are those that yield information on the location of pixels and land cover classes relative to each other. It is important to remember that p_i is often used directly in the calculation of several metrics configurations and thus can covertly or overtly govern much of the "behavior" of these spatially explicit metrics. The simple fact that some metrics appear to be tightly linked to proportion both through correlations (Neel et al., 2004) and the fact that p_i is often used directly in the formulae for many metrics (Turner et al., 2001) makes several metrics of configuration potentially problematic for use in assessing fragmentation. Correlation coefficients between many metrics and proportion may exceed 0.90, and correlations between many metrics and level of aggregation may exceed 0.80 (Neel et al., 2004).

Interestingly, an important dilemma in empirical field studies of organismal response to fragmentation is controlling for and distinguishing the impact of fragmentation from the *independent* effects of habitat loss (or proportion) (Fahrig, 2003; Harrison and Bruna, 1999). This distinction is also important in the realm of watershed management that attempts to link land cover change to changes in water quality (Gergel, 2005). Given the close linkages between proportion and configuration pattern indices, it is not surprising that dissecting the independent effects of habitat loss (a nonspatial process) from habitat fragmentation (an inherently spatially explicit process) has been quite difficult. Such questions are among the core of our understanding for quantitatively linking pattern and process. Thus, metrics that are strongly related to aggregation but independent of class area are the most robust and useful metrics to quantify fragmentation (Neel et al., 2004).

Neighbor Rules, Patch Definition, and Landscape Connectivity

Questions regarding spatial configuration often involve calculating the total number of patches, mean patch size, or total amount of edge in a landscape. All of these metrics require information about how individual pixels are located relative to their "neighbors," and all require some assumptions on how to determine what a neighbor actually is. Neighbor rules refer to the definition one uses to determine pixel and patch connectivity. For example, when two pixels of the same class share a flat edge and are considered part of the same patch; this is referred to as the four-neighbor rule. The eight-neighbor rule is used when two pixels of the same class share at least one corner and are considered connected and part of the same patch. Not surprisingly, estimates of landscape change can vary drastically depending on the neighbor rule (Gardner, 2001). No comparisons of landscape fragmentation through time should *ever* be reported without explicit mention of the neighbor rule used in the analysis. For example, a landscape could appear to become more fragmented in 2000 (as compared to 1980) simply because a four-neighbor rule for defining connectivity was used in 2000 and an eight-neighbor rule was used in 1980.

The choice of neighbor rule is often more of an art than a science but should always be governed by the questions under consideration. Movement patterns of insects, for example, vary by flying ability (Driscoll and Weir, 2005) and habitat (Ross et al., 2005; Schooley and Wiens, 2004), whereas movement patterns of plants may depend on the dispersal agent (e.g., birds, cattle, or seed rain) (Arrieta and Suarez, 2005). In practice, an eight-neighbor rule is often used to define hydrologically connected landscapes and to define the connectivity among adjacent wetlands. Eight-neighbor rules are commonly used when defining adjacencies for good dispersers (e.g., birds) or for processes such as fire that can jump as they spread across a landscape. Connectivity for organisms and processes that disperse poorly (unable to cross gaps or travel far distances) are often mapped using four-neighbor rules.

Last, any discussion of connectivity should distinguish between structural connectivity and functional connectivity. Structural connectivity (which pixels are adjacent to others) is easily quantified, down to several significant digits, using the variety of software packages described next. Functional connectivity is in fact more important and more relevant to management and conservation. Functional connectivity may be dynamic and variable for any process or organism and is much more difficult to quantify and map. For example, the gap-crossing ability of flying insects may render patches several pixels away as functionally "connected" when individuals are able to traverse 100 m of generally inhospitable habitat to reach the next suitable patch of habitat. The spread of fire might also be looked at in this context as the ability to "jump" from one fire-prone area to the next and is influenced by the direction and slope of spread (e.g., up- or downslope) and may change due to changing wind speed and direction. Understanding these mechanisms that influence how structural connectivity translates into functional connectivity for the spread of insects, pathogens, fire, and the like, is key to translating static landscape patterns into the dynamic landscape processes that are ultimately of interest.

TOOLS OF THE TRADE

Once the above issues regarding the input data set and goals of the analysis and questions are formalized, there exist a variety of software packages to perform the pattern analysis. Among the most widely used and known is *Fragstats* (McGarigal et al., 2002), originally released to the public in 1995 as part of research conducted by the U.S. Department of Agriculture Forest Service and has since become the industry standard. This software exemplifies not only the enormous utility that can be derived from such software, but also can easily lead the uninformed user to make some beginner's mistakes. Some of these common mistakes include:

1. Not providing a clear rationale or theoretical reason for why the metrics were chosen (Gustafson, 1998). At a minimum, this involves a hypothesis for the expected direction of change in a metric. Ideally, this involves a likely explanation for the ecological reasons and mechanisms causing this change.
2. Including results from too many metrics. Many metrics are highly correlated, so the presentation of more than a defined small set is rarely warranted. Techniques such as principal components analysis can help reduce the redundancy of metrics.
3. Not considering and matching the scale of analysis to the scale of the question at hand (Gustafson, 1998). If possible, one should try to conduct the analysis at different scales.

The importance of these points cannot be emphasized enough. Neglect of these points has substantially limited our ability to link pattern metrics with process, the ultimate goal of pattern analysis. In addition to *Fragstats*, other freely available software has been developed, and these packages vary in their utility for research versus direct management applications (Table 7.2). Along with conducting pattern analyses on an image, several programs include the ability to create simulated maps, the importance of which is discussed in the Real and Artificial Landscapes section. Furthermore, at http://www.csiss.org/ clearinghouse/ a wide variety of additional

TABLE 7.2
Additional Freely Available Software Packages for Calculation of Landscape Metrics or Landscape Pattern Indices (LPIs)

Software	Description	Reference
Apack 2.17	Similar in many respects to *Fragstats* but generally faster. The program originated as a linkage to the Landis forest disturbance model. *Apack* has since been redesigned by others with a specific emphasis on handling thousands of larger data sets in a computationally efficient manner and can now take advantage of multiple processors. The software requires an *Erdas* GIS or ASCII text format for input files.	Mladenoff and DeZonia (1999)
ClaraT	Emphasizes the diversity of ways to calculate and analyze fractals. The software can be used to explore the concepts of self-similarity, random growth processes, and stochastic renormalization. Exercises have been developed for *ClaraT* that examine self-affine functions and linkages to time series analysis (Milne et al. 1999).	Milne et al. (1999)
Leap II	The Landscape Ecological Analysis Package was designed for management-related purposes, such as the monitoring of spatial patterns through time in response to differing management and policy options. The program was designed for use with spatial fire regime simulators and interfaces with *Fragstats*, *ArcView*, *Arc/Info*, *Erdas* are in various stages of development.	Perera et al. (1997)
Rule	Calculates landscape metrics but was originally designed for the creation of simulated, highly controllable replicated maps, primarily for research purposes. Fragmentation and proportion can be varied independently using two simple parameters. The software includes area-weighted metrics and the option for lacunarity analysis, but does not overwhelm the user with an enormous number of pattern indices. Educational materials have been developed to assist users (Gardner 1999, 2001).	Gardner (1999)
Simmap 2.0	Runs in a PC Windows environment and computes several landscape pattern indices related to edges, number, size, and shape of the patterns and others. Raster output can be saved as images easily importable to other programs. The program also generates realistic simulated maps using algorithms different from *Rule*. The relative abundance of cover classes and fragmentation can be varied independently and systematically controlled. Simulations are generally rapid, only a few seconds for an 800×800 map.	Saura and Martinez-Millan (2000)

programs may be found. Selection of an appropriate program can be determined by the specific concern, with discipline-specific programs for spatial questions related to crime hot spots, mining, epidemiology, and diverse social science applications.

There are also extensions created to work within a GIS platform to conduct spatial analyses. The major (nontrivial) benefit of such extensions is that they avoid the creation and export of ASCII grids or text files from the GIS for import into a separate self-contained program. This is important as files are often huge and unwieldy. This can become quite problematic when multiple landscapes are analyzed, particularly if one does not have access to automated code for repeated analyses of many maps. Researchers in such a situation may particularly benefit from programs such as the *Patch Analyst ArcView Extension* (Rempel and Carr, 2003), or r.le (Baker and Cai, 1992), which is essentially a series of commands written for *GRASS GIS* (open source GIS software).

NEW CONCEPTS OF FRAGMENTATION

Traditional ways of viewing and measuring habitat fragmentation have largely been influenced by the Island Biogeography perspective (MacArthur and Wilson, 1967), in which habitat fragments are seen as islands within an inhospitable matrix. The pioneer of plant ecology John Curtis's classic map of landscape change through time in Cadiz Township, Wisconsin (Figure 7.1), represents this quintessential view of habitat fragmentation from this island perspective. As this perspective has been so influential in shaping conservationists' view of reserves and reserve design (Diamond, 1976), it is not surprising that many of the most widely used metrics for assessing habitat fragmentation include the numbers of patches and mean patch size, as the relevance of such metrics for a "Curtis-style" fragmented landscape is fairly obvious (Figure 7.1). The aforementioned problem of dissecting the independent effects of nonspatial habitat loss from spatially explicit habitat fragmentation is also evident from such a map.

However, in other parts of the world, habitat fragmentation is occurring at different rates and through different causes, leading to a variety of fragmented landscapes that do not fit this island model. Several types and phases of habitat fragmentation have been identified (Forman, 1995; Jaeger, 2000; McIntyre and Hobbs, 1999). One specific typology of landscape fragmentation introduced the idea of variegated landscapes (McIntyre and Hobbs, 1999) that occur along a continuum from intact to variegated, fragmented, and relictual landscapes (Figure 7.2a). The concept of a variegated landscape may be more relevant in largely intact landscapes, typical of many regions of Canada, for example. Deforestation is among the most widely quantified of the different land cover conversion processes. Thus, it has become clear that land cover conversion is often not uniform or random but tends to be clustered (Lepers et al., 2005), often in highly productive areas and often at the edge of forested areas and along roads. Another useful categorization therefore includes four main types of fragmentation that occur as a result of different spatial processes: linear, insular, diffuse, and massive fragmentation (Jeanjean et al., 1995). Figure 7.2b shows additional phases of fragmentation that have been identified,

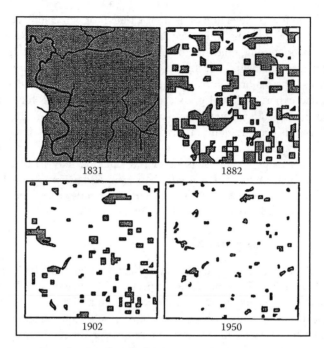

FIGURE 7.1 Classic example of habitat fragmentation from Cadiz Township, Wisconsin, by John Curtis as a result of European settlement. The dark shaded areas represent intact habitat, with clear patterns of habitat fragmentation of remaining patches from 1831 to 1882 to 1902 to 1950. (Originally from Curtis, J., in W. L. Thomas (Ed.), *Man's Role in Changing the Face of the Earth*, University of Chicago Press, Chicago, 1956.)

including perforation and incision by linear agents such as roads, as well as dissection, dissipation, shrinkage, and attrition (Forman, 1995; Jaeger, 2000) (Figure 7.2b).

Rather than debate specific terminology of the classifications, note several points of interest (Figure 7.2). First, note the very different dominant patterns occurring in early versus later stages of the process, as well as the diversity of causes (e.g., linear vs. more diffuse). It is likely that not all metrics behave consistently across all phases of fragmentation or are useful in all types of landscapes. Thus, it is important to distinguish among landscape pattern indices that are useful in the early versus later stages of fragmentation, as well as those more or less sensitive to linear drivers of fragmentation. The basic landscape pattern indices of use in an island-type landscape (e.g., number of patches, mean patch size) might not always be of use in other situations. "Perforated" landscapes might respond particularly well to different metrics of fragmentation, such as lacunarity, that attempt to address the "holiness" of landscapes. These distinctions are quite relevant to consider depending on whether one's focal class of interest is widely dissected across the landscape or makes up the background matrix of the landscape. As important as determining the correct metric to describe the structural connectivity of a landscape, it ultimately is more important to appropriately represent the functional connectivity of a landscape throughout the process of fragmentation of the landscape as the point at which a

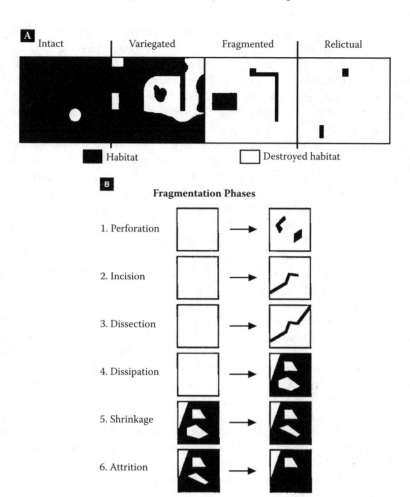

FIGURE 7.2 Classification of landscapes according to various stages of fragmentation. Panel A shows a scheme proposed by McIntyre and Hobbs (1999) that includes four general categories based on the degree of fragmentation. Panel B presents the phases of the fragmentation process (from Jaeger, 2000, and modified and extended after Forman, 1995, p. 407, Figure 12.1). Note the initial and ending patterns are similar to the original patterns in the classic Curtis example, but the intermediary causes and patterns are quite distinct: the first phases of fragmentation are driven largely by linear disturbances as well as the perforated landscape patterns.

landscape becomes functionally fragmented varies greatly by process or organism (Neel et al., 2004).

ADDITIONAL MEASURES OF FRAGMENTATION

New landscape metrics are under continuous development. This is driven in part by a need to better quantify functional connectivity of landscapes (as some metrics are

better correlated with dispersal success Schumaker, 1996). Due to the extreme importance of the measuring forest fragmentation (and its inverse, habitat connectivity) and
given that a variety of new metrics have been proposed to improve on traditional
metrics of connectivity, several of the recent new ideas in measuring habitat fragmentation are discussed next. This discussion includes several indices not routinely calculated by all software packages (Table 7.1). As a result, they may not be as familiar
in the literature. This list is certainly not exhaustive, however, discussed are several
new metrics that may be particularly suited to address questions previously unanswer-
able by the "standard metrics." This idea is revisited at the end of this chapter within
the context of new approaches and new metrics needed for fine-scale pattern analysis.

Three recent LPIs, landscape division, effective mesh size, and splitting index,
have been devised by Jaeger (2000) in an attempt to better link fragmentation indices
to ecological processes of interest. Very loosely, these three metrics are meant to
relate to the likelihood of two organisms finding each other (or not) on a given
landscape. Landscape division is the probability that two random locations are both
situated in the cover class of interest. While all three are closely related, they are
different in their interpretation and behavior. In a comparison of these three newer
metrics to a suite of more commonly used LPIs, many traditional LPIs were deemed
useful for distinguishing only certain phases of fragmentation but not others (Jaeger,
2000), whereas, the new proposed metrics were useful for all phases of fragmentation.

Furthermore, a comparison that involved degree of landscape division, splitting
index, effective mesh size (as well as two road density measures), found that only
these three new metrics responded consistently to increasing fragmentation. Metrics
such as number of patches and average patch size were inconsistent (Jaeger, 2000)
and should be replaced by effective mesh size. Effective mesh size is also recommended when one must compare among landscapes of differing extent. As a result
of this work, useful criteria for evaluating the utility of fragmentation metrics have
been proposed and include intuitive interpretation, low sensitivity to extremely small
patches, monotonous changes to different stage of fragmentation, and mathematical
simplicity (Jaeger, 2000).

Lacunarity (Mandelbrot, 1983) is a measure of texture and has been used within
the context of quantifying landscape-level patterns (Plotnick et al., 1993). *Lacunarity*
refers to the "holiness" or "gappiness" of a landscape. Lower lacunarity occurs for
patterns that contain similar gap sizes, whereas higher lacunarity occurs in landscapes
with a wider range of gap sizes (Plotnick et al., 1993). Unlike measures such as
contagion (O'Neill et al., 1988), which is merely a fine-scale measure of texture,
lacunarity allows the examination of multiscale texture (Plotnick et al., 1993). This
is important as fragmentation can occur at finer as well as broader scales. Lacunarity
is the result of a moving window (or gliding box) analysis described further in the
work of Plotnick et al. (1993) and Gardner (1999). A lacunarity analysis can provide
information about the overall portion of the map occupied by the cover class of interest,
the contagion at a given scale, the scale at which map patterns are random, and the
scales over which the landscape exhibits self-similar patterns (Plotnick et al., 1993).

It is important to remember that lacunarity is a scale-dependent measure; therefore, both high and low values can be observed on the same landscape when
measured at different scales. As such, a measurement of lacunarity at one scale, or

the use of one single value, is generally not extremely useful. Instead, the power of lacunarity is the information provided by measurements made at different scales (Plotnick et al., 1993). Furthermore, in contrast to many other neighborhood analysis techniques, measurements made at multiple scales are not sensitive to map edges. However, lower lacunarity values (as the box size increases) are likely to be detected (Plotnick et al., 1993). Lacunarity is also reliable for use on sparse maps, which may make it useful to consider further in fragmentation studies. This metric deserves more examination for variegated landscape patterns and for perforating agents of landscape change, as in Figure 7.2b.

Interestingly, the Matheron index (Matheron, 1970), essentially a core/perimeter ratio, deserves mention as it has been used by the remote sensing community for quite some time but has not been used at all in the landscape ecology literature. Some interesting work showing the behavior of the Matheron index (Matheron, 1970) is reproduced here in Figure 7.3 (Achard et al., 2001). In this case, the index was calculated for every 9×9 pixel block on a coarse spatial resolution map (AVHRR image). Note that the range of fragmentation is greatest at 50% forest cover. Thus, the behavior of fragmentation indices must be evaluated within the context of the remaining proportion of habitat on the landscape in case of mathematical artifacts related to correlations with p_i, as well as due to the diversity of conceptualizations of the phases of fragmentation previously discussed.

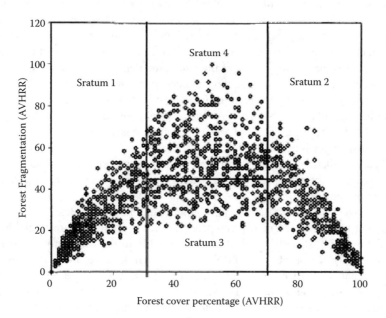

FIGURE 7.3 Relationship between percentage cover of forest versus one measure of forest fragmentation (Achard et al., 2001). The amount of fragmentation varies with proportion of forests, as does the variability in fragmentation values. Shown here is the Matheron index (Matheron, 1970) of fragmentation.

EDGES AND ECOTONES

An *ecotone* is an area where the majority of the variables (species or environmental factors) show the highest rates of change (Burrough, 1986). Some have further distinguished an *edge* as referring to only sharper changes, while using the term *boundary* to refer to both sharp and more gradual changes (Fagan et al., 2003). Both sharp edges and gradual boundaries can influence the exchange of materials or indicate that a flux of material or energy is occurring in an area. Edges are quite important ecologically given the increased diversity, disturbance, microclimate conditions, exotic species, and other effects often observed at edges. Defining edges is also critical to defining patches and thus defining our view of habitat fragmentation. In most metric calculation packages, edge can be quantified in a variety of ways using different specific algorithms (e.g., *Rule* vs. *Fragstats Algorithms*) and reported in different units, such as the length of linear edge (kilometers) or total area of edge habitat (hectares).

Edges are also, unfortunately, among the most misclassified portions of a landscape image. Classification errors can be due to a variety of factors, but mixed pixels (where the field of view includes more than one cover type) (Hlavka and Livingston, 1997) are particularly problematic. Edge errors become more important as the landscape becomes more fragmented with many edges between cover classes. This is particularly ironic as such landscapes are precisely those landscapes which a fragmentation analysis is of most interest to managers, conservation biologists, etc. One option uses "area-weighted" and "core area" metrics. Such metrics help mitigate edge problem as they essentially "ignore" the portions of the landscape where classification errors are highest. The legitimacy and implications of this approach, which ignores the "edgey" areas of the landscape that are often of most interest in a fragmentation analysis, are troubling and deserving of more attention. Thus, when characterizing forest fragmentation, edge-related forest fragmentation measures might be particularly problematic. As such, special consideration should be paid to the study of edges through all phases of a pattern analysis research, from classification to specifications made when using *Fragstats*.

Many of the advancements in edge detection have occurred outside the realm of the standard software packages for pattern analysis. One group of edge detection algorithms requires evenly spaced data, thus rendering them useful for remotely sensed imagery, and is based on wombling (Womble, 1951). Briefly, on a grid, a wombling algorithm essentially compares a point to its four nearest neighbors by calculating the first partial derivative of a variable in the four cardinal directions (Fortin, 1994). A wombling "surface" is formed that describes the magnitude of the rates of change observed for given variables; the surface is level when the rate of change is zero (or very low) and inclined when the variable shows a higher rate of change. A boundary or ecotone is defined as those areas where wombling values are high (above some chosen threshold) (Fortin, 1994). Values can also be averaged among several variables to obtain a wombling surface for a combination of variables. Such techniques, however, will almost *always* detect a gradient of some sort. Therefore, significance tests are essential in determining if the observed rates of change are different from what might be observed at random. Interestingly, testing for significance using a simple null random pattern will lead to a conservative test compared to the use of a null pattern with spatial autocorrelation.

Boundaries detected for different types of vegetation structure in deciduous forest in Quebec using wombling yielded interesting results (Fortin, 1997). Delineated boundaries were compared using trees only, shrubs only, as well as trees and shrubs combined and involved comparisons using densities, percentage cover, and presence/absence data (Fortin, 1997). Overlap among boundaries was determined using an overlap statistic: O_s. In many cases, similar boundaries were found regardless of whether density, percentage cover, or presence/absence was used. One useful aspect of this work for remote sensing is that boundaries detected using tree percentage cover corresponded well to boundaries defined using densities, possibly fortuitous for ground validation, for which swiftly and easily collected measurements are desirable (Fortin, 1997). The generality of this, especially when applied to new high spatial resolution imagery sources, would be interesting to test.

IMPORTANT CONSIDERATIONS FOR PATTERN ANALYSIS OF FOREST DISTURBANCE

AVOID REDUNDANT METRICS

While a seemingly amazing array of landscape metrics is possible to calculate, it is clear that many are quite highly correlated and thus redundant (Griffith et al., 2000; Neel et al., 2004; Riitters et al., 1995). As such, the number of metrics reported in any study will generally need to be reduced substantively from the entire set output from the software. The number of metrics can be reduced through the careful selection of a small set of metrics clearly related to a specific hypothesis (Langford et al., 2006) or reduced in number via principal components analysis, factor analysis, or other multivariate reduction techniques (Cain et al., 1997; Cardille et al., 2005; Cumming and Vernier, 2002; Imbernon and Branthomme, 2001; Riitters et al., 1995).

There have been several indications that there may be four or five distinct dimensions that can measured on a landscape using landscape metrics (Cain et al., 1997; Cumming and Vernier, 2002; Li and Reynolds, 1995; Riitters et al., 1995). These general groupings may even be robust to changes in grain (Griffith et al., 2000). One factor analysis (Cain et al., 1997) found fairly consistent relationships across different data sets. Primarily, the "texture factor" was the most important in all data sets and could explain the majority of variance regardless of spatial resolution and other factors (Cain et al., 1997). This analysis suggested four general groupings for metrics. The first group of metrics included texture variables such as land cover dominance and contagion. The second group included patch shape and compaction, and this grouping was not particularly robust. A third group included fractal estimators of perimeter complexity, and the fourth group included only the number of classes.

Even fewer groupings of LPIs may be necessary (Neel et al., 2004). A thorough analysis of *Fragstats* output calculated from a variety of simulated replicate landscapes yielded three general behavioral groupings of metrics: those that are closely correlated with proportion or areal measures (or composition), a second group closely tied to the spatial autocorrelation (or configuration), and a last group with less-predictable behavior (including unstable, nonlinear, threshold responses) in response to proportion and autocorrelation (Neel et al., 2004). The second class of metrics, those that respond to aggregation independent of proportion, are among the best suited for assessing habitat fragmentation. A detailed table with class-level LPIs, associated behavioral groupings, and correlations with proportion and autocorrelation is presented in the work of Neel et al. (2004). Examination of this table by all readers is highly recommended. Furthermore, the nonlinear behavior of many metrics suggests linear analytical techniques may be inappropriate (Neel et al., 2004). Unless there is an ecological or geophysical reason to calculate many metrics, there is little need for it (Cain et al., 1997). Thus, it is generally appropriate to examine the correlations among any metrics one might consider using and then only report a limited number of metrics.

CONTEXT NEEDED WHEN QUANTIFYING DISTURBANCE PATTERNS

Characterizing fragmentation and natural disturbance regimes invariably requires some measure of central tendency to describe a situation. However, in many cases the distribution of pattern metrics is nonnormal, possibly log-normal, rendering mean values problematic for adequately summarizing pattern. Examining the distribution of patch sizes will generally be much more illuminating than only examining the mean patch size. For example, mean fire size, as determined by a "mean patch size" metric for burned areas or the average size of patches estimated by *Fragstats*, might not actually ever occur on a landscape. Instead, median values or, preferably, a frequency distribution, are better suited for describing actual patterns. Furthermore, the sensitivity of any measure of central tendency to the characteristics of the input data (such as grain and extent) should also be addressed explicitly before patterns can be accurately quantified. However, few researchers have the time or capability to fully address patterns in such detail in every study.

Furthermore, when examining the distribution of patch sizes, small patches can be quite important to quantify as in some cases the distribution of patch sizes is dominated by many very small patches. For example, one study examined the statistical distribution of patch sizes for burn scars in tropical savannah using Landsat Multi-Spectral Scanner (MSS) and ponds in the Arctic using ERS-1 synthetic aperture radar images (Hlavka and Livingston, 1997), two very fragmented cover types. They determined that 24,776 of 27,698 ponds were less than 0.0028 km^2 and 472 of 477 total burn scars were less than 8 km^2. Last, their comparison of the distribution of burn scars between AVHRR and Landsat revealed three types of distortion with the coarse spatial resolution imagery (Hlavka and Livingston, 1997). The study stressed that while patches smaller than a pixel cannot be detected, patches of equal size to a pixel might likely be detected in neighboring pixels (Hlavka and Livingston, 1997). Clearly, the possibility of small patches approaching the size of the MMU or grain size of the image must be carefully considered. Other features of the map itself will greatly influence the perceived patch size distribution as truncation of the map effectively limits the size of the largest patch, while the resolution and MMU will have an impact on the smallest patch size visible.

Patch size distributions of fragmented cover classes can be modeled with theoretical distributions (Hlavka and Livingston, 1997), and these distributions are not necessarily normally distributed. An analysis of patch size distributions that compared the fit of a Pareto (power) distribution and an exponential distribution resulted in r^2 values ranging from 79 to 99% for these distributions (Hlavka and Livingston, 1997). Comparing mean values of patch sizes, particularly from distributions that are not the same, is potentially extremely problematic. This could be a problem for other LPIs as well. For example, some results suggest that mean shape index in particular should not be used when comparing maps with different patch size distributions (Saura, 2002).

Diverse tools are needed to adequately describe the natural and unnatural range of variability in landscape pattern and provide the context for the results (Cumming and Vernier, 2002; Cushman and Wallin, 2000; Kemper et al., 2000; Staus et al., 2002; Tinker et al., 2003). In addition to empirical data from individual studies attempting to characterize natural and anthropogenic sources of variability in landscape disturbance, two sources of information are essential if landscape ecologists and others want truly to understand the broader context in which their particular landscape patterns reside. Next, a brief overview of these two additional approaches is provided. These approaches can aid specifically in the aforementioned problems of varying patch size distributions among different landscapes, interactions between proportion and arrangement, and the impact of grain, extent, and MMU.

WHY THE INTERPLAY BETWEEN PATTERN ANALYSIS IN "REAL" AND "ARTIFICIAL" LANDSCAPES IS ESSENTIAL

While diverse definitions of spatial heterogeneity abound, there is generally no agreed-on quantitative definition of landscape heterogeneity. As a result, teasing out variability in pattern caused by natural versus anthropogenic forces still remains problematic (as discussed in Chapter 8, this volume). Ultimately, most pattern analyses fundamentally ask questions about the causes and consequences of spatial heterogeneity. An important goal of pattern analysis should be to quantify landscape patterns rigorously to determine which landscapes are outside or approaching the limits of their natural range of variability. Certainly, a more quantitative definition of landscape heterogeneity might be useful for comparing among different landscapes and asking: Which landscapes are more heterogeneous than others? How did this heterogeneity arise? What portion of the heterogeneity is due to human causes versus natural sources of variability?

Simply comparing among a few landscapes or contrasting one landscape over multiple time periods might indeed detect differences, but whether the changes observed are ecologically or statistically significant is generally unknowable with small sample sizes. Meaningful interpretation of the actual relevance of an LPI is difficult without some context for its actual range and variability in the real world. An obvious example might be examining 20 years of landscape change using the available archive of Landsat images in a region where the dominant organism lives well beyond 20 years. This might not provide the entire view of natural levels of variation across such a landscape.

Thus, new ways of quantifying the range in metrics that move beyond comparing two or three landscapes are needed. Such methods would use a multitude of replicated landscapes that span the possible range of variation due to natural sources as well as the possible range of variation due to anthropogenic causes. The first of these databases, called META-LAND, has been developed in the United States (Cardille et al., 2005), originating in the agriculture-dominated midwestern regions of the United States but rapidly growing to include new areas, including Canada, starting with British Columbia (Gergel et al., in preparation). This system offers the unprecedented ability to select from the full range of natural variability actually seen in landscapes. With such databases, pattern analysis of one landscape through time will have the necessary context to determine if the changes actually seen are out of the ordinary. Furthermore, metric behavior on "real landscapes" cannot be understood until there is also an understanding of how landscape metrics behave on artificially created maps with only controlled sources of variability.

NEUTRAL LANDSCAPE MODELS

An essential key to understanding the behavior of landscape metrics is to understand and interpret their behavior in highly controlled situations on landscapes that have been designed (and replicated) to vary in explicit ways, an experimental design largely unobtainable in the field (Langford et al., 2006; Neel et al., 2004; Saura and Martinez-Millan, 2000). These simulated maps have been termed neutral landscape models (NLMs) (Gardner et al., 1987) and spatial stochastic models (Fortin et al., 2003).

Such simulated maps are an important complement to the use of classified imagery for evaluating LPIs for a variety of important reasons. So-called real landscapes — maps and classified remotely sensed imagery — have errors. Some of these errors are unknown, unquantifiable, and biased. Patterns seen in real landscapes may be biased due to natural sources of variability or postprocessing and data collection methods. For example, anisotropy can have an impact on the value and range of variability seen in LPIs, and one may not even be aware of the magnitude and direction of such a bias in an image. Creating artificial landscapes enables control of such sources of bias, creating highly controlled replicate landscapes that are difficult to find in the real world. NLMs enable one to ask many of the questions posed in this chapter while controlling for the sources of variability *that* are not of interest and explicitly manipulating the sources of variability that are of interest (e.g., p_i and aggregation; see Figure 7.4). Neutral landscape models may also be quite useful for the above-mentioned "chicken-and-egg" problem by which the amount of cover class affects the resulting fragmentation metric derived from a landscape, and the measured area of a land cover class can be impacted by the fragmentation patterns seen on the ground.

Major advances in realism have been made in the latest generation of neutral landscape models (Table 7.2) using fractal midpoint displacement algorithms (Gardner et al., 1987) and the modified random clusters method (Saura and Martinez-Millan, 2000). The realism of artificial maps is improving such that, in a recent survey involving over 100 map experts, the respondents were unable to discern synthetic maps from actual maps (Hargrove et al., 2002). Thus, quite realistic,

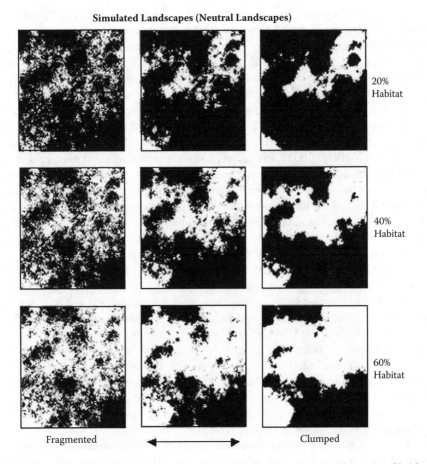

FIGURE 7.4 Several examples of simulated neutral landscapes showing the range of habitat proportions as well as fragmentation that can be systematically manipulated. Both can be varied independently in programs such as *Rule* (Gardner, 1999) and *SimMap* (Saura and Martinez-Millan, 2000). The ability to distinguish the effects of habitat loss independently from the changes due to habitat fragmentation is a crucial challenge in pattern analysis and can be aided by the use of simulated maps. These landscapes were originally used in the work of Langford et al., 2006.

multiple-category maps are possible with independently controllable categories such as clumpiness and proportion (Gardner, 1999; Remmel and Csillag, 2003) as both have an impact on the expected values and the variance in LPIs on simulated maps (Gardner, 1999). Landscapes with linear, noodlelike features and nonstationary anisotropy (Gardner, 1999; Hargrove et al., 2002) have also been created. To the author's knowledge, all of the current options for simulated maps (and all reported data on their behavior) are based solely on rectangular landscapes, the boundaries of which bear little resemblance to the shapes of natural physiographic, ecoregional, watershed-level, or geomorphic boundaries often seen in actual landscapes, aside from agricultural fields.

NLMs have been fundamental to the development of a vast array of theory in landscape ecology (With and King, 1997). While simulated maps may seem only useful in the realm of theory and of little value for management and on-the-ground decision making, it has become apparent that to generate a full range of behaviors from landscapes, improvements in NLMs are essential. For example, if the behavior of a new (or existing) landscape pattern index has only been evaluated in one location, its behavior remains largely unknown. Further, if the expected value and variance of a metric is not known, there is really no basis on which to decide what constitutes significant landscape change (or significant difference) (Csillag and Boots, 2005; Remmel and Csillag, 2003). Instead, an LPI should be evaluated relative to some underlying statistical distribution of the LPI (Fortin et al., 2003), which is difficult to quantify when $N = 1$ or $N = 2$ landscapes. Thus, simulating the probability distribution of an LPI, ideally over a range of proportions and magnitudes of autocorrelation (Fortin et al., 2003), is an essential part of devising a rigorous statistical test to detect statistically significant landscape change. In general, the statistical distribution over which most metrics occur is not well understood except in the case of random landscapes. Despite this, LPI values are often compared among landscapes without reference to any underlying distribution.

The goal of creating better spatial models for null hypothesis testing is not lost on other disciplines (Goovaerts and Jacquez, 2005). Efforts are under way from organizations such as the U.S. National Cancer Institute to fund research into neutral models for use in pattern recognition for health and epidemiological purposes. Interestingly, this was motivated in part by a need to address the problems and bias toward false positives that can be created by unrealistic and oversimplified null hypotheses such as complete spatial randomness. It is likely that expansion of new and improved neutral spatial model software will continue as it is explored in disciplines beyond that of landscape ecology.

In summary, it is clear that comparing any two LPIs devoid of any context for the probability distribution from which they arise is nonilluminating and may border on nonsensical. However, it is also important to point out that there are benefits and disadvantages to comparing the value of an LPI to a probability distribution derived from images of actual landscapes (the real world) versus a probability distribution derived from simulated maps. Probability distributions derived from real maps are fraught with unknown and unquantifiable errors and biases. Simulated maps benefit from the ability to span the entire theoretical (and likely wider) range of possible values consistently for a given LPI through systematic control and replication. However, the ecological significance of an LPI value that falls within or outside the theoretical distribution may be harder to determine as the simulated probability distribution may span a much greater range of variability than actually observed for an LPI on real landscapes. Thus, statistically significant and ecologically significant differences in LPIs must both be addressed.

CHALLENGES AND OPPORTUNITIES

A variety of future challenges faces the users and creators of spatial data used in pattern analysis. These challenges span the range of purely conceptual to entirely technical, as well as challenges that will involve both. Due to the earlier assertion that changes in the discipline of remote sensing science have been the driving force behind the development of landscape pattern analysis, the focus here is on issues that appear promising for active collaboration between the two disciplines and that yield practical benefits to those using pattern analysis for decision making.

ACCURACY ASSESSMENT OF LANDSCAPE METRICS

As any map typically contains error, any analysis derived from that map will also contain errors. There are diverse potential causes of errors in an image that can lead to over- or underestimation of the areal amounts of different cover classes (Achard et al., 2001). These causes of errors include data compositing procedures, data resampling or rescaling, classification techniques, training sites, and number of classes used (Wulder, Boots, et al., 2004). For example, errors as high as 40–50% have been reported for areal estimates of burns (Setzer and Pereira, 1991). One of the primary ways in which errors in a map might be related to errors in landscape metrics is through inaccuracies in the classification. However, research on the accuracy of landscape metrics is generally sparse (Langford et al., 2006).

Early work suggested little need for concern regarding the impact of classification accuracy upon resulting landscape metrics. Wickham et al. (1997) suggested that if differences in land cover composition were roughly 5% greater than the misclassification rate, it could be assumed that the calculated differences between LPIs were not due to misclassification. Recent work has reported mixed conclusions about the severity of the issue (Brown et al., 2000; Hess and Bay, 1997; Shao et al., 2001; Wickham et al., 1997). Of the limited number of studies that have explicitly addressed the impact of classification error on propagation of errors to metrics, the ability to generalize the findings remains limited.

Furthermore, caution should be exercised as studies based on "real maps" have inherent biases and trends. Often these maps have also been subjected to a MMU correction as well. Thus, extrapolation of any quantitative prediction of metric error from these studies is problematic. The results only apply to a given range of maps and structural types of landscapes as examined in the studies. Issues related to map accuracy and the relationship to errors in landscape metrics is particularly well-suited to the aforementioned combined approaches of "bulk" pattern analysis and neutral landscape models that are necessary to have large sample sizes of replicate landscapes that truly span the diversity of landscape proportions and structure. As a consequence, these same studies could be built upon by spanning a greater range of proportions, spatial autocorrelation, and fragmentation.

Intuitively, it may be thought that a more accurately classified map would result in more accurate landscape metrics. In one of the more systematic attempts to compare map classification accuracy and landscape metrics accuracy using thousands of simulated maps, replicated with systematically controlled proportions and aggregation of land cover), it was found that percentage error of the metrics did not relate well to the percentage classification error on the same map (Langford et al., 2006). The study focused on the landscape metrics routinely used to quantify habitat fragmentation, and the impact on other metrics remains to be examined.

As mentioned, errors in the areal extent of different cover classes on a classified image are often greater as the landscape becomes more fragmented with many edges between cover classes (Roesch et al., 1995). Pattern (as measured using spatial autocorrelation) was seen early in the remote sensing literature as having an impact on accuracy assessment (Congalton, 1988). This is particularly troubling as it is precisely in such fragmented landscapes that one is most likely interested in conducting a pattern analysis for management or conservation purposes.

The problem of inaccuracies in landscape metrics when measured across multiple scales might also be exacerbated by the fragmented structure of landscapes. For example, several studies have examined the correlations between fine-scale data sources and coarser scale data sets (Jeanjean and Achard, 1997). Interestingly, the correlations between deforestation measured using Landsat TM and coarser-scale AVHRR data sources were different depending on the degree of fragmentation (Jeanjean and Achard, 1997). In landscapes with high levels of fragmentation $r^2 = 0.36$ and areas with lower levels of fragmentation yielded r^2 values of 0.85. Thus, the discrepancy (or errors) among the two data sets were clearly of greater concern in the fragmented situation, whereas agreement among the different data sources (thus corroborating the accuracy of both) was higher when the landscape was largely intact. Such errors in the areal measures of cover classes and the extent to which these errors translate into metric errors remain to be determined.

The recurring (and somewhat circular) chicken-and-egg problem between proportional areal measures and fragmentation measures is troubling. It is clear that the proportion of the landscape occupied by a cover class has an enormous impact on any resulting configuration metrics that may be calculated for that landscape. However, the original landscape structure, particularly when fragmented, can have an impact on the resulting areal proportion measures on an image. Neutral landscape models can help with this problem through an attempt to control for these factors separately or *a priori*, but it is clear this circularity must be addressed. Because we are aware that classification errors are generally higher at patch edges, core area metrics may be useful in some instances (where the patch edges are ignored and only the core area of the patches are used by *Fragstats* to calculate metrics). Given the importance of edges to a variety of management and conservation concerns (e.g., negative edge effects), this solution may not be feasible for all applications.

Accuracy assessment of imagery is an important research issue and practical challenge for both map users and producers. While specifics are routinely debated, there nonetheless exists an entire body of literature, accepted methods, jargon, and tools associated with accuracy assessment (Foody, 2002). There exists no such parallel for accuracy assessment of landscape metrics. In addition, much of the traditional, standardized ways of addressing accuracy assessment of imagery are generally nonspatial and of limited value for directly assessing the validity of any pattern analysis. User and producer errors do not give any indication of where in the landscape errors exist. In addition, reporting overall accuracies of land cover classifications (Cain et al., 1997) rather than per class estimates, is problematic, especially for rare cover classes, as the low number of reference points renders the confusion matrix problematic (Langford et al., 1996). In one example, "background" matrix cells had both producer's and user's accuracy exceeding 99%, while the more rare fire class had producer's and user's accuracies of just over 12% and 7%, respectively, with a resulting overall accuracy exceeding 99%. This is particularly problematic when the rare cover types (e.g., remaining habitat, small patchy disturbances) are those classes subject to a pattern analysis.

Few studies have examined errors in landscape metrics, and each has resulted in different conclusions. From the few studies that do exist, there is limited ability to generalize to the broad range of applications undertaken with pattern analysis, much less to make any quantitative predictions regarding the nature or magnitude of errors in landscape metrics that may be a result of errors in base maps. The limited results that exist, however, suggest that much more research, and more systematic research, is needed to answer this question (Langford et al., 2006). This oversight in research, the lack of our ability to assess the direction and magnitude of errors in landscape metrics, could have potentially serious consequences for research, management, and policy as there is likely error of unknown magnitude in the results of every pattern analysis ever conducted (Langford et al., 2006). Furthermore, our assessments of natural levels of variability in landscape pattern could be unduly influenced by the variability in metrics simply caused by error. This will inhibit our ability to prescribe management treatments (i.e., based on some ideal number or size of burns) or to quantify unnatural levels of insect outbreak at the landscape scale.

PATTERN ANALYSIS USING HIGH SPATIAL RESOLUTION IMAGERY

The vast majority of published research involving landscape pattern analysis has used Landsat TM data. Reliable, repeat availability (since the 1970s) at low prices (or free) has rendered it the true "workhorse" of landscape ecology. The importance of the repeated imagery, allowing the tracking of changes over time, has revolutionized the way ecology, global ecology, and many other disciplines see the world. Thus, most of our ideas about landscape pattern analysis are derived from a Landsat view of the world. Questions such as "How has landscape X changed through time?" are often implicitly asking, "How have the land cover classes associated with 30 × 30 m pixels changed through time?" Issues with the continuity of the Landsat series of satellites are not explored further here (some details may be found in Chapter 2 of this volume).

New high spatial resolution sensors are under development and deployment, resulting in data made available more cheaply and readily (e.g., QuickBird, IKONOS). Thus, scientists have an unprecedented opportunity to envision landscape patterns composed of meter and even submeter size pixels. The new research questions empowered by such approaches are numerous, as are the new challenges (Wulder, Hall, et al., 2004). The obvious challenge will be reconcilation with the common Landsat view of the world: many cover types per pixel versus the high-resolution perspective with many pixels per cover type. The major errors in landscape metrics previously discussed (mixed pixels at edges of cover types) may become less relevant as newer problems arise (e.g., finer scale patterns and problems associated with shading and structure of the understory).

Are current metrics and approaches relevant for pattern analysis on finer-resolution imagery? How does clumpiness (spatial autocorrelation) vary with scale as we move from a worldview with many objects per pixel to one with many pixels per object? An interesting first attempt at linking pattern metrics with fine-scale measurements was in tropical forests in Sabah, Malaysia (Marsden et al., 2002). The goal was to quantify fine-scale structure of understory vegetation at a scale relevant to a foraging animal. Lateral photographs of the understory vegetation were converted to black-and-white images, scanned, and then read into *Fragstats* for a pattern analysis (Marsden et al., 2002). Findings of the study included that density measures were intercorrelated (such as mean patch size and number of patches), and complexity measures (shape and fractal dimension) were also intercorrelated. However, the shape measures were not significantly correlated with the complexity measures (Marsden et al., 2002). This study suggested that pattern metrics traditionally used at broader spatial scales may also be useful for vegetation at fine scales. It also suggested that a coherent set of metrics can be used to measure shape consistently, while a different set may be useful for measuring complexity. More important, the work suggested that these measures of shape and complexity are quantifying distinct, uncorrelated aspects of vegetation structure.

Furthermore, as finer-scale data are often used as "truth" for accuracy assessment of a coarser-scale image (Achard et al., 2001; Lepers et al., 2005), it is important to revisit some of the potential issues with this approach. Correlations between a fine- and coarse-scale image may be influenced by fragmentation levels (Jeanjean and Achard, 1997). Recall, the results of regression analysis of percentage cover measured at a fine resolution (e.g., Landsat TM) versus percentage cover measured at a more coarse spatial resolution (e.g., AVHRR) differed in fragmented and continuous landscapes (Achard et al., 2001). It was suggested that it may be useful to control for the effect of this variable (fragmentation) on the variance in the parameter of interest (area in forest cover) (Achard et al., 2001) by selecting sites based on several strata that account for the relationship between percentage deforestation and forest fragmentation. Referring again to their results in Figure 7.3, it is suggested that a selection of ground truth sites occur in each of four strata: first stratum is low proportion of forest (<30%) and low fragmentation index; second stratum is 30–50% forest and low fragmentation index; third stratum is proportion ranging from 50 to 70% and high fragmentation index; and a fourth stratum is more than 70% proportion and high fragmentation index. Use of 36 sites selected in this way resulted in a strong $r^2 = 0.94$ for percentage forest using Landsat TM versus percentage cover using National Oceanic and Atmospheric Administration AVHRR imagery. For the highly fragmented second stratum, the Matheron index may also be used in a correction procedure (Achard et al., 2001) to control for the oft-reported spatial aggregation bias in the coarser-scale AVHRR (as well as other possible sources of bias and error in the imagery) (Hlavka and Livingston, 1997; Jeanjean and Achard, 1997). Surrogates for fragmentation, such as population density, have also been suggested as useful correction factors for adjusting AVHRR estimates of percentage forest with finer-scale data sources (Roesch et al., 1995). The "independence" of such surrogate fragmentation measures from any bias in the particular image from which it is obtained will be interesting to explore further.

High spatial resolution remote sensing has the potential to fundamentally alter our view of, and questions posed of, landscapes, and will likely influence the relevant scale over which many pattern analyses are conducted. It may influence the call for more continuous categorizations to represent forest cover classes (e.g., varying by tree cover) (Lepers et al., 2005) as opposed to traditional discrete categorical land cover classes.

LINEAR LANDSCAPE METRICS

As mentioned, among the different fragmentation types, such as insular, diffuse, massive, and linear (Jeanjean et al., 1995), linear landscape metrics are chronically absent from most software calculation packages and are not in routine use. This is likely in part due to the nature of raster data and associated software. However, linear features are far more important in governing landscape pattern than the general lack of attention in pattern analysis would suggest.

The impact of roads on ecological processes are diverse and substantive (Switalski et al., 2004) and may extend for a considerable distance beyond the road itself (Forman, 1999; Hansen and Clevenger, 2005). Incorporation of roads into fragmentation analyses has been debated (Riitters et al., 2004); however, for some landscape processes, the impact of roads is greater than the impact of other land cover changes. For example, in a pattern analysis in the Rocky Mountains, the amount of edge habitat created by roads was 1.54–1.98 times greater than that resulting from harvest (Reed et al., 1996). The effect of roads is particularly striking when accounting for the limited area actually occupied by roads on the landscape (Guthrie, 2002) as drainage patterns in watersheds can be fundamentally rerouted by roads, altering flows and possibly increasing the risk of landslides (Guthrie, 2002; Roberts et al., 2004; Tague and Band, 2001). Both paved and unpaved roads are among the drivers of further landscape change, with some of the best examples from the Amazon (Kirby et al., 2006). Roads can result in injury and mortality to native organisms or act as barriers to dispersal. Roadside vegetation may help foster the spread of invasive species (Hansen and Clevenger, 2005). Other linear disturbances abound, such as ski runs (Laiolo and Rolando, 2005) and seismic lines for energy exploration (Bayne et al., 2005), which would likely benefit from descriptions other than "mean road density" that provide little context for the location of road expansion in a landscape.

Rivers and riparian zones are another smaller cover type, often highly fragmented, that could benefit from improved linear LPIs as well as the use of high spatial resolution remote sensing technology. In many regions of the world, riparian habitat types are highly linear — they may only occupy 5% of the landscape along a river corridor but provide important habitat to the majority of the species in a region (especially arid environments). Riparian areas are also subject to multiple disturbances that tend to differ from those in the uplands (e.g., floods, scouring, deposition), and when riparian buffers or riparian management zones are left on the landscape as part of best management practices, such areas might be particularly susceptible to blowdown. Thus, the characterization of disturbance patterns in these areas forms an integral part of truly defining the natural range of variability in a watershed. Historically, these areas have also been notoriously hard to map using remotely sensed data at 30-m cell sizes and have been problematic in a variety of studies. Thus, new high-resolution data sources should be considered a "revolution" for mapping riparian areas.

Characterizing and quantifying linear patterns and disturbances on the landscape are clearly important. Even if traditional LPIs can capture most of the structural fragmentation of a landscape without explicitly examining roads (Riitters et al., 2004), the unique impacts of linear features on ecosystems and functional connectivity may warrant them separate consideration. While road densities can easily be determined, such measures are largely divorced from the surrounding landscape context. Such summary measures do not enable quantification of such questions as how many roads traverse a small stream or bisect a large reserve. As finer spatial resolution imagery will enable better distinctions among primary, secondary, and tertiary roads, the ability to understand this fundamental driver of landscape change will undoubtedly improve markedly. That said, linear metrics would be expected to be *quite* susceptible to classification error as narrow linear features could be shattered and split by the mere misclassification of one or a few pixels. For all of these reasons, linear features deserve further attention in pattern analysis.

CONCLUSIONS

Pattern analysis is a rapidly expanding area of study that has the potential to benefit diverse disciplines in environmental science. First, a series of heuristics was relayed to aid in the formulation of the "nuts and bolts" of a pattern analysis. Second, the results of pattern analyses were discussed more broadly within the context of quantifying natural levels of variability in forest pattern and how this area of research can be best guided by, and benefit from, the tools of pattern analysis. Concomitant with this expansion is the potential for misuse. Some components of a rigorous pattern analysis include:

1. Clearly stated questions and hypotheses
2. A reduction in the total number of metrics reported as many metrics are redundant
3. An examination of the frequency distributions of LPIs, not just mean values
4. Acknowledgment that most metrics are "wrong" in some way, with an effort to identify the most obvious sources of errors and bias
5. Consideration of the accuracy of edge pixels, which may be particularly misclassified

Understanding the theoretical bounds as well as actual behavior of LPIs remains a challenge (Neel et al., 2004). It is clear, however, that proportion (composition) and autocorrelation (configuration) have strong, and sometimes interactive and nonlinear influence on LPI values. Comparison of LPIs from landscapes with no information on the probability distribution from which the LPI arose and without explicit consideration of proportion and autocorrelation can yield less-rigorous conclusions. Continued research using diverse techniques is needed to quantify natural and anthropogenic variability in LPIs. This should include empirical work supplemented by analysis of landscapes "in bulk" (using many replicates) and involve both actual and synthetic maps to truly capture the widest range of possible landscape structures. Identification of statistically significant and ecological significant differences in LPIs is important, and both must be addressed. Furthermore, creation and application of metrics that quantify functional connectivity, not just structural connectivity, are essential, as are metrics that capture the diverse types and stages of the fragmentation process.

As advances in remote sensing have directly and indirectly driven much of the evolution of the techniques, future research by the remote sensing community has the potential to greatly improve the quantitative rigor of pattern analysis. Several specific knowledge gaps in pattern analysis that could be aided by collaborations with the remote sensing community include assessing the interplay between errors in imagery and the metrics derived from them, the development of landscape metrics that are relevant for use with high-resolution imagery, and improved methods for quantifying and mapping linear landscape feature disturbances. It is through such continued collaborations that the best technical and conceptual advances in both fields can aid our ability to quantify and, it is hoped, manage the dynamic changes currently undergoing in the landscapes of the world.

ACKNOWLEDGMENTS

The thoughts in this chapter were influenced greatly by my discussions and collaborations with remote sensing scientists, particularly William Langford, Jeff Cardille, and Nicholas Coops.

REFERENCES

Achard, F., Eva, H., and Mayaux, P. (2001). Tropical forest mapping from coarse spatial resolution satellite data: production and accuracy assessment issues. *International Journal of Remote Sensing*, 22, 2741–2762.

Arrieta, S. and Suarez, F. (2005). Spatial dynamics of *Ilex aquifolium* populations seed dispersal and seed bank: understanding the first steps of regeneration. *Plant Ecology*, 177, 237–248.

Baker, W.L. and Cai, Y. (1992). The r.le programs for multi-scale analysis of landscape structure using the GRASS geographical information system. *Landscape Ecology*, 7, 291–302.

Bayne, E.M., Van Wilgenburg, S.L., Boutin, S., and Hobson, K.A. (2005). Modeling and field-testing of Ovenbird (*Seiurus aurocapillus*) responses to boreal forest dissection by energy sector development at multiple spatial scales. *Landscape Ecology*, 20, 203–216.

Benson, B.J. and MacKenzie, M.D. (1995). Effects of sensor spatial resolution on landscape structure parameters. *Landscape Ecology*, 10, 113–120.

Brown, D.G., Duh, J.D., and Drzyzga, S.A. (2000). Estimating error in an analysis of forest fragmentation change using North American Landscape Characterization (NALC) data. *Remote Sensing of Environment*, 71, 106–117.

Burrough, P.A. 1986. Principles of Geographic Information Systems for Land Resources Assessment. Oxford University Press. Oxford, U.K. 50pp.

Cain, D.H., Riitters, K., and Orvis, K. (1997). A multi-scale analysis of landscape statistics. *Landscape Ecology, 12,* 199–212.

Cardille, J.A. and Turner, M.G. (Eds.), (2001). *Understanding Landscape Metrics I.* Springer-Verlag, New York. 316p.

Cardille, J.A., Turner, M.G., Clayton, M., Gergel, S.E., and Price, S. (2005). METALAND: a publicly available tool for characterizing spatial patterns and statistical context of landscape metrics across large areas. *Bioscience,* 55(11):983–988.

Congalton, R.G. (1988). Using spatial autocorrelation analysis to explore the errors in maps generated from remotely sensed data. *Photogrammetric Engineering and Remote Sensing, 54,* 587–592.

Csillag, F. and Boots, B. (2005). A framework for statistical inferential decisions in spatial pattern analysis. *The Canadian Geographer, 49,*172–179.

Cumming, S. and Vernier, P. (2002). Statistical models of landscape pattern metrics, with applications to regional scale dynamic forest simulations. *Landscape Ecology, 17,* 433–444.

Curtis, J. (1956). The modification of mid-latitude grasslands and forests by man. In W. L. Thomas (Ed.), *Man's Role in Changing the Face of the Earth.* University of Chicago Press, Chicago. 1236p.

Cushman, S.A. and Wallin, D.O. (2000). Rates and patterns of landscape change in the Central Sikhote-alin Mountains, Russian Far East. *Landscape Ecology, 15,* 643–659.

Diamond, J.M. (1976). Island biogeography and conservation — strategy and limitations. *Science, 193,* 1027–1029.

Driscoll, D.A. and Weir, T. (2005). Beetle responses to habitat fragmentation depend on ecological traits, habitat condition, and remnant size. *Conservation Biology, 19,* 182–194.

Fagan, W.F., Fortin, M.J., and Soykan, C. (2003). Integrating edge detection and dynamic modeling in quantitative analyses of ecological boundaries. *Bioscience, 53,* 730–738.

Fahrig, L. (2003). Effects of habitat fragmentation on biodiversity. *Annual Review of Ecology Evolution and Systematics, 34,* 487–515.

Foody, G.M. (2002). Status of land cover classification accuracy assessment. *Remote Sensing of Environment, 80,* 185–201.

Forman, R.T.T. (1995). *Land Mosaics: The Ecology of Landscapes.* Cambridge University Press, Cambridge, U.K. 652p.

Forman, R.T.T. (1999). Horizontal processes, roads, suburbs, societal objectives and landscape ecology. In J.M. Klopatek and R.H. Gardner (Eds.), *Landscape Ecological Analysis: Issues and Applications.* Springer-Verlag, New York. 400p.

Fortin, M.J. (1994). Edge-detection algorithms for two-dimensional ecological data. *Ecology, 75,* 956–965.

Fortin, M.J. (1997). Effects of data types on vegetation boundary delineation. *Canadian Journal of Forest Research, 27,* 1851–1858.

Fortin, M.J., Boots, B., Csillag, F., and Remmel, T.K. (2003). On the role of spatial stochastic models in understanding landscape indices in ecology. *Oikos, 102,* 203–212.

Gardner, R.H. (Ed.). (1999). *RULE: Map generation and a Spatial Analysis Program.* Springer-Verlag, New York. 400p.

Gardner, R.H., (Ed.). (2001). *Neutral Landscape Models.* Springer-Verlag, New York. 316p.

Gardner, R.H., Milne, B.T., Turner, M.G., and O'Neill, R.V. (1987). Neutral models for the analysis of broad-scale landscape pattern. *Landscape Ecology, 1,* 19–28.

Gergel, S.E. (2005). Spatial and non-spatial factors: when do they affect landscape indicators of watershed loading? *Landscape Ecology, 20,* 177–189.

Gergel, S.E., D' Eon, R.G., and Cardille, J.A. (In preparation). Natural levels of variability at the watershed-scale.

Goovaerts, P. and Jacquez, G.M. (2005). Detection of temporal changes in the spatial distribution of cancer rates using local Moran's I and geostatistically simulated spatial neutral models. *Journal of Geographical Systems*, *7*, 137–159.

Greenberg, J.D., Gergel, S.E., and Turner, M.G. (Eds.). (2001). *Understanding Landscape Metrics II: Effects of Changes in Scale*. Springer-Verlag, New York. 316p.

Griffith, J.A., Martinko, E.A., and Price, K.P. (2000). Landscape structure analysis of Kansas at three scales. *Landscape and Urban Planning*, *52*, 45–61.

Gustafson, E.J. (1998). Quantifying landscape spatial pattern: what is the state of the art? *Ecosystems*, *1*, 143–156.

Guthrie, R.H. (2002). The effects of logging on frequency and distribution of landslides in three watersheds on Vancouver Island, British Columbia. *Geomorphology*, *43*, 273–292.

Hansen, M.J. and Clevenger, A.P. (2005). The influence of disturbance and habitat on the presence of non-native plant species along transport corridors. *Biological Conservation*, *125*, 249–259.

Hargrove, W.W., Hoffman, F.M., and Schwartz, P.M. (2002). A fractal landscape realizer for generating synthetic maps. *Conservation Ecology*, *6*, 2.

Harrison, S. and Bruna, E. (1999). Habitat fragmentation and large-scale conservation: what do we know for sure? *Ecography*, *22*, 225–232.

Hess, G.R. and Bay, J.M. (1997). Generating confidence intervals for composition-based landscape indexes. *Landscape Ecology*, *12*, 309–320.

Hlavka, C.A. and Livingston, G.P. (1997). Statistical models of fragmented land cover and the effect of coarse spatial resolution on the estimation of area with satellite sensor imagery. *International Journal of Remote Sensing*, *18*, 2253–2259.

Imbernon, J. and Branthomme, A. (2001). Characterization of landscape patterns of deforestation in tropical rain forests. *International Journal of Remote Sensing*, *22*, 1753–1765.

Jaeger, J.A. (2000). Landscape division, splitting index, and effective mesh size: new measures of landscape fragmentation. *Landscape Ecology*, *15*, 115–130.

Jeanjean, H. and Achard, F. (1997). A new approach for tropical forest area monitoring using multiple spatial resolution satellite sensor imagery. *International Journal of Remote Sensing*, *18*, 2455–2461.

Jeanjean H., Fonts J., Puig H. and Husson, A. (1995). Study of forest non-forest interface typology of fragmentation of tropical forest. In *Tropical Ecosystem Environment Observation by Satellite (TREES) Series B: Research Reports No. 2*. European Commission Joint Research Centre and European Space Agency, Luxembourg. EUR — 16291EN. 90p.

Kemper, J., Cowling, R.M., Richardson, D.M., Forsyth, G.G., and McKelly, D.H. (2000). Landscape fragmentation in South Coast Renosterveld, South Africa, in relation to rainfall and topography. *Austral Ecology*, *25*, 179–186.

Kirby, K.R., Laurance, W.F., Albernaz, A.K., Schroth, G., Fearnside, P.M., Bergen, S., Venticinque, E.M., and. da Costa, C. (2006). The future of deforestation in the Brazilian Amazon. *Futures,* 38(4):432–453.

Kleinn, C. (2000). Estimating metrics of forest spatial pattern from large area forest inventory cluster samples. *Forest Science*, *46*, 548–557.

Laiolo, P. and Rolando, A. (2005). Forest bird diversity and ski-runs: a case of negative edge effect. *Animal Conservation*, *8*, 9–16.

Landres, P.B., Morgan, P., and Swanson, F.J. (1999). Overview of the use of natural variability concepts in managing ecological systems. *Ecological Applications*, *9*, 1179–1188.

Langford, W., Gergel, S.E., Dietterich, T.G., and Cohen W. (2006). Map misclassification can cause large errors in landscape pattern indices: examples from habitat fragmentation. *Ecosystems*, *9*, 1–16.

Lepers, E., Lambin, E.F., Janetos, A.C., DeFries, R., Achard, F., Ramankutty, N., and Scholes, R.J. (2005). A synthesis of information on rapid land-cover change for the period 1981–2000. *Bioscience*, *55*, 115–124.

Li, H. and Reynolds, J.F. (1995). On definition and quantification of heterogeneity. *Oikos*, *73*, 280–284.

MacArthur, R.H. and Wilson, E.O. (1967). *The Theory of Island Biogeography*. Princeton University Press, Princeton, NJ. 224p.

Mandelbrot, B.B. (1983). *The Fractal Geometry of Nature*. W.H. Freeman, New York. 480p.

Marsden, S.J., Fielding, A.H., Mead, C., and Hussin, M.Z. (2002). A technique for measuring the density and complexity of understorey vegetation in tropical forests. *Forest Ecology and Management*. *165*, 117–123.

Matheron, G. (1970). *La theorie des variables regionalisees et ses applications*. Fascicule 5 des cahiers du Centre de Morphologie Mathematique de Fontainebleau, France. 212p.

McGarigal, K., Cushman, S.A., Neel, M.C., and Ene. E. (2002). *Fragstats*: Spatial Pattern Analysis Program for Categorical Maps. Computer software program produced by the authors at the University of Massachusetts, Amherst, USA. Available at http://www.umass.edu/landeco/research/fragstats/fragstats.html.

McIntyre, S. and Hobbs, R. (1999). A framework for conceptualizing human effects on landscapes and its relevance to management and research models. *Conservation Biology*, *13*, 1282–1292.

Milne, B.T., Johnson, A.R., and Matyk, S. (Eds.), (1999). *ClaraT: Instructional Software for Fractal Pattern Generation and Analysis*. Springer-Verlag, New York. 400p.

Mladenoff, D. and DeZonia, B. (1999). *APACK 2.17 Analysis Software User's Guide*. Retrieved from http://landscape.forest.wisc.edu/projects/APACK/apack.html on Sept. 15, 2005.

Neel, M., McGarigal, C.K., and Cushman, S.A.(2004). Behavior of class-level landscape metrics across gradients of class aggregation and area. *Landscape Ecology*, *19*, 435–455.

O'Neill, R.V., Hunsaker, C.T., Timmins, S.P., Jackson, B.L., Jones, K.B., Riitters, K.H., and Wickham, J.D. (1996). Scale problems in reporting landscape pattern at the regional scale. *Landscape Ecology*, *11*, 169–180.

O'Neill, R.V., Krummel, J.R., Gardner, R.H., Sugihara, G., Jackson, B., DeAngelis, D.L., Milne, B.T., Turner, M.G., Zygmont, B., Christensen, S.W., Dale, V.H., and Graham, R.L. (1988). Indices of landscape pattern. *Landscape Ecology*, *1*, 153–162.

Perera, A.J., Baldwin, D.J.B., and Schnekenburger, F. (1997). *LEAP II: A Landscape Ecological Analysis Package for Land Use Planners and Managers*. Ontario Ministry of Natural Resources, Ontario Forest Research Institute, Sault Ste. Marie, ON, Canada.

Plotnick, R.E., Gardner, R.H., and O'Neill, R.V. (1993). Lacunarity indices as measures of landscape texture. *Landscape Ecology*, *8*, 201–211.

Reed, R.A., Johnson-Barnard, J., and Baker, W.L. (1996). Contribution of roads to forest fragmentation in the Rocky Mountains. *Conservation Biology*, *10*, 1098–1106.

Remmel, T.K. and Csillag, F. (2003). When are two landscape pattern indices significantly different? *Journal of Geographical Systems*, *5*, 331–351.

Rempel, R.S. and Carr, A.P. (2003). *Patch Analyst extension for ArcView*, version 3, Centre for Northern Forest Ecosystem Research, Lakehead, Thunder Bay, Ontario, Canada.

Riitters, K., Wickham, J., and Coulston, J. (2004). Use of road maps in national assessments of forest fragmentation in the United States. *Ecology and Society*, *9*, 13. Retrieved from http://www.ecologyand society.org/vol9/iss2/art13/ on Sept. 15, 2005.

Riitters, K.H., O'Neill, R.V., Hunsaker, C.T., Wickham, J.D., Yankee, D.H., Timmins, S.P., Jones, K.B., and Jackson, B.L. (1995). A factor-analysis of landscape pattern and structure metrics. *Landscape Ecology*, *10*, 23–39.

Roberts, B., Ward, B., and Rollerson., T. (2004). A comparison of landslide rates following helicopter and conventional cable-based clear-cut logging operations in the Southwest Coast Mountains of British Columbia. *Geomorphology*, *61*, 337–346.

Roesch, F.A., Vandeusen, P.C., and Zhu, Z.L. (1995). A comparison of various estimators for updating forest area coverage using AVHRR and forest inventory data. *Photogrammetric Engineering and Remote Sensing*, *61*, 307–311.

Ross, J.A., Matter, S.F., and Roland, J. (2005). Edge avoidance and movement of the butterfly *Parnassius smintheus* in matrix and non-matrix habitat. *Landscape Ecology*, *20*, 127–135.

Saura, S. (2002). Effects of minimum mapping unit on land cover data spatial configuration and composition. *International Journal of Remote Sensing*, *23*, 4853–4880.

Saura, S. (2004). Effects of remote sensor spatial resolution and data aggregation on selected fragmentation indices. *Landscape Ecology*, *19*, 197–209.

Saura, S. and Martinez-Millan, J. (2000). Landscape patterns simulation with a modified random clusters method. *Landscape Ecology*, *15*, 661–678.

Schooley, R.L. and Wiens, J.A. (2004). Movements of cactus bugs: patch transfers, matrix resistance, and edge permeability. *Landscape Ecology*, *19*, 801–810.

Schumaker, N.H. (1996). Using landscape indices to predict habitat connectivity. *Ecology*, *77*, 1210–1225.

Setzer, A.W. and Pereira, M.C. (1991). Amozonia biomass burnings in 1987 and an estimate of their tropospheric emissions. *Ambio*, *20*, 19–22.

Shao, G., Liu, D., and Zhao, G. (2001). Relationships of image classification accuracy and variation of landscape statistics. *Canadian Journal of Remote Sensing*, *27*, 35–45.

Shen, W.J., Jenerette, G.D., Wu, J.G., and Gardner, R.H. (2004). Evaluating empirical scaling relations of pattern metrics with simulated landscapes. *Ecography*, *27*, 459–469.

Staus, N.L., Strittholt, J.R., DellaSala, D.A., and Robinson, R. (2002). Rate and pattern of forest disturbance in the Klamath-Siskiyou ecoregion, U.S.A. between 1972 and 1992. *Landscape Ecology*, *17*, 455–470.

Switalski, T.A., Bissonette, J.A., DeLuca, T.H., Luce, C.H., and Madej, M.A. (2004). Benefits and impacts of road removal. *Frontiers in Ecology and the Environment*, *2*, 21–28.

Tague, C. and Band, L. (2001). Simulating the impact of road construction and forest harvesting on hydrologic response. *Earth Surface Processes and Landforms*, *26*, 135–151.

Tinker, D.B., Romme, W.H., and Despain, D.G. (2003). Historic range of variability in landscape structure in subalpine forests of the Greater Yellowstone Area, U.S.A. *Landscape Ecology*, *18*, 427–439.

Turner, M.G., Gardner, R.H., and O'Neill, R.V. (2001). *Landscape Ecology in Theory and Practice*. Springer-Verlag, New York. 401p.

Wickham, J.D., Oneill, R.V., Riitters, K.H., Wade, T.G., and Jones, K.B. (1997). Sensitivity of selected landscape pattern metrics to land-cover misclassification and differences in land-cover composition. *Photogrammetric Engineering and Remote Sensing*, *63*, 397–402.

Wickham, J.D. and Riitters, K.H. (1995). Sensitivity of landscape metrics to pixel size. *International Journal of Remote Sensing*, *16*, 3585–3594.

With, K.A. and King, A.W. (1997). The use and misuse of neutral landscape models in ecology. *Oikos*, *79*, 219–229.

Womble, W.H. (1951). Differential systematics. *Science*, *114*, 315–322.

Wu, J.G. (2004). Effects of changing scale on landscape pattern analysis: scaling relations. *Landscape Ecology*, *19*, 125–138.

Wulder, M.A., Boots, B., Seemann, D., and White, J. (2004). Map comparison using spatial autocorrelation: an example using AVHRR derived land cover of Canada. *Canadian Journal of Remote Sensing*, 30, 573–592.

Wulder, M.A., Hall, R.J., Coops, N.C., and Franklin, S.E. (2004). High spatial resolution remotely sensed data for ecosystem characterization. *Bioscience*, *54*, 511–521.

8 Characterizing Stand-Replacing Harvest and Fire Disturbance Patches in a Forested Landscape: A Case Study from Cooney Ridge, Montana

*Andrew T. Hudak, Penelope Morgan,
Mike Bobbitt, and Leigh Lentile*

CONTENTS

INTRODUCTION

In this chapter, we present a case study intended to help crystallize for many readers, through use of an illustrative example, some of the important concepts developed in the preceding chapters. From an understanding of forest successional and disturbance processes, both natural and anthropogenic (Linke et al., Chapter 1, this volume), research questions were developed to compare and contrast the landscape patterns generated from fire and harvest disturbance. Remotely sensed data are demonstrated as an appropriate source of relevant information (Coops et al., Chapter 2, this volume), enabling the applications presented for the utilization of change detection approaches for mapping of forest harvest (Healey et al., Chapter 3, this volume) and fire (Clark and Bobbe, Chapter 5, this volume). As presented in Chapter 6 (Rogan and Miller, this volume), the use of supportive spatial data sets to aid in the analysis and interpretation of the maps and patterns exhibited is demonstrated. The forest harvest and fire maps are subjected to pattern analysis as outlined by Gergel (Chapter 7, this volume), providing insights into the research questions identified.

FOREST HARVEST AND FIRE DISTURBANCES

Timber harvest and fire are influential disturbance processes affecting many forested landscapes in the American West. These forests are managed for a variety of human values, including residential, recreational, wildlife habitat, water quality, and wood production purposes. If managers are to mimic the effects of natural disturbances, then they must integrate the timing and severity of prescribed disturbances with the ecological requirements of the desired landscape composition and condition. Understanding the effects of different types of disturbances and associated alteration of key processes may help to promote ecosystem resiliency through improved management decisions (Kimmins, 1997). Both forest and fire management practices influence succession, and the individual and cumulative effects of disturbances may have positive and negative implications for ecosystem character and function (Moore et al., 1999; Tinker and Baker, 2000). Development of sustainable relationships between humans and their environments requires knowledge of successional consequences.

Forest harvests vary in extent and intensity, but some degree of change in soil and water properties and loss of nutrients will occur in any harvested system (Pritchett and Fisher, 1987). In general, clearcutting alters microclimatic, soil, vegetation, animal habitat, and microbial conditions more severely than less-intensive or partial cutting. Clearcutting favors early successional microclimates and tolerable levels of vegetation competition but may not create the type of forest floor environment conducive to regeneration of desired species (Kimmins, 1997). High surface temperatures and low surface soil moisture content may lead to slow revegetation rates following clearcutting, although invasive species may find these conditions favorable (Pritchett and Fisher, 1987). Forest harvest, especially clearcutting, may have greater influence in hot and dry climates or on steep slopes where the potential for soil erosion and slow rates of plant recovery is high, particularly if timber harvesting is coupled with other intensive practices such as grazing, repeat burning, and farming (Smith et al., 1996). Furthermore, clearcutting results in fragmented forests with altered age, structural, and spatial characteristics, which may have important implications for wildlife habitat, bird nesting success, and landscape diversity (Mladenoff et al., 1993; Tinker and Baker, 2000).

The term *burn severity* is broadly defined as the degree of ecosystem change induced by fire and encompasses fire effects on both vegetation and surface soils (Key and Benson, in press; Ryan, 2002; Ryan and Noste, 1985). Severe fires are those that result in great ecological changes (De Bano et al., 1998; Johnson et al., 2003; Moreno and Oeschel, 1989; Rowe, 1983; Ryan, 2002; Ryan and Noste, 1985; Schimmel and Granstrom, 1996). If "severity" is considered a relative term, then severe fires are so named because they slow vegetation recovery, alter nutrient cycles, or increase abundance of invasive species, tree mortality, or soil erosion potential to an undesirable, perhaps even unnatural, degree. The short-term effects of recent severe fires have been studied (Graham, 2003; Lewis et al., 2006; Turner et al., 1997), but there remains limited understanding of the longer-term effects of severe fires on forest demography and structure (Savage and Nystrom Mast, 2005).

Burn severity varies greatly at fine scales in Africa (Brockett et al., 2001), North America (Hudak, Morgan, et al., 2004; Hudak, Robichaud, et al., 2004), and elsewhere, but the causes and consequences of that spatial variability in terms of postfire effects are poorly understood. Recent developments in remote sensing and vegetation pattern analysis allow the evaluation of burn severity, which influences subsequent vegetation recovery (White et al., 1996). The degree to which prior timber harvest and other vegetation conditions have influenced fire effects across landscapes is little understood yet has tremendous implications for the efficacy of fuel management designed to moderate fire effects.

OBJECTIVE AND ANALYSIS APPROACH

Our objective is to demonstrate consistent and objective use of remote sensing and geographical information system (GIS) tools to characterize and compare the patch characteristics of stand-replacing harvest and fire disturbance processes in a coniferous forest landscape where both disturbances were known to have recently occurred. Consistency and objectivity are required for conducting a reliable remote sensing analysis in the absence of explicit ground validation data (Hudak and Brockett, 2004; Hudak, Fairbanks, et al., 2004), as was the case in this study. We do, however, have substantial and sufficient local knowledge of the Cooney Ridge area and wildfire event to conduct this study.

The two satellite-based spectral indices applied in this analysis were the middle-infrared corrected normalized difference vegetation index (NDVIc) (Nemani et al., 1993) and the normalized burn ratio (NBR) (Key and Benson, in press). Pocewicz et al. (2004), working in mixed-conifer forest in the northern Rocky Mountains, found NDVIc to be a better predictor of leaf area index than the more broadly applied, uncorrected NDVI; "correcting" the NDVI with a middle-infrared band increased the sensitivity of the index to forest biomass. Therefore, we selected NDVIc to indicate forest biomass.

Key and Benson (in press) found that NBR outperformed NDVI as a predictor of composite burn index, an integrated, ecological field measure of burn severity based on vegetation and soil effects, due to the higher sensitivity of NBR to soil effects. As a result, NBR is the burn severity index used by the U.S. Forest Service Remote Sensing Applications Center (RSAC) and the U.S. Geological Survey Earth Resources Observation and Science Data Center; both produce burned area reflectance classification (BARC) maps to inform rapid response Burned Area Emergency Rehabilitation (BAER) team decisions on large, active wildfire events (as described in Clark and Bobbe, Chapter 5, this volume). We also selected NBR to indicate burn severity in this study, but it must be noted that both NDVI and NBR are more sensitive to green vegetation cover than to the underlying soils (Hudak, Morgan, et al., 2004; Hudak, Robichaud, et al., 2004). Therefore, in this study we consider "severe fire" to be more indicative of a lack of green vegetation cover than to any soil effects.

Because we wished to map forest cover *change* as a result of stand-replacing harvest and fire disturbances rather than simply forest cover *condition*, whenever possible we employed image-differencing techniques (delta, d) to indicate forest harvest with dNDVIc and fire-induced vegetation mortality with dNBR.

METHODS

STUDY AREA

The study area (20,672 Ha) is topographically rugged, with elevations ranging from 1129 to 2353 m (Figure 8.1 and Figure 8.2). Vegetation is mixed-conifer forest type, with the important conifer species *Pseudotsuga menziesii* (Douglas fir), *Larix occidentalis* (western larch), *Pinus contorta* (lodgepole pine), *Pinus ponderosa* (ponderosa pine), *Abies lasiocarpa* (subalpine fir), and *Picea engelmannii* (Engelmann spruce) (A. Hudak, 2003). Common shrubs are *Physocarpus malvaceus* (mallow ninebark), *Alnus incana* (thinleaf alder), *Symphoricarpos albus* (common snowberry), *Rubus parviflorus* (thimbleberry), *Shepherdia canadensis* (russet buffaloberry), *Vaccinium membranaceum* (thinleaf huckleberry), *Spiraea betulifolia* (birchleaf spirea), *Mahonia repens* (creeping barberry), *Acer glabrum* var. *douglasii* (Rocky Mountain maple), *Lonicera utahensis* (Utah honeysuckle), and *Rosa* spp. (rose). Common forbs include *Chamerion angustifolium* (fireweed), *Arnica cordifolia* (heartleaf arnica), *Apocynum androsaemifolium* (spreading dogbane), *Linnaea borealis* (twinflower), and *Xerophyllum tenax* (common beargrass). Common grasses include *Calamagrostis rubescens* (pinegrass), *Festuca idahoensis* (Idaho fescue), *Phleum pratense* (timothy), *Agrostis scabra* (ticklegrass), and *Elymus glaucus* (blue wildrye) (L. Lentile, 2004). *Equisetum* spp. (horsetail) and *Peltigera aphthosa* (freckle pelt lichen) commonly occur. *Centaurea maculosa* (spotted knapweed), a

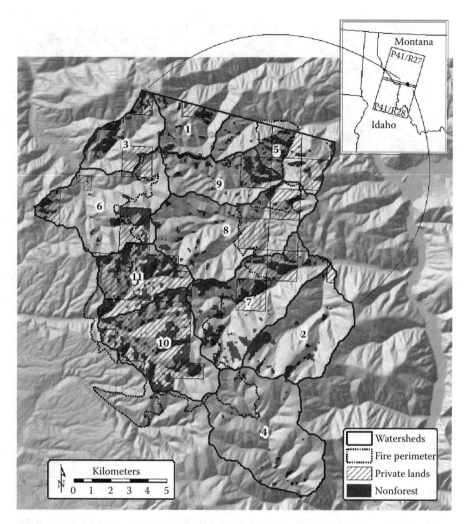

FIGURE 8.1 Shaded relief of the Cooney Ridge study area in relation to the relevant Landsat Path/Row footprints. The 11 watersheds defining the study area (20,672 Ha) are delineated and ranked in ascending order according to proportion within the 26 August 2003 wildfire perimeter. Lands within the study area not indicated as private or nonforest are public forest lands.

Category I noxious weed in Montana, frequents roadsides and other disturbed areas. Forest habitat types in the study area range from warm, dry *P. menziesii* habitat types that support fire-maintained *P. ponderosa*, to cooler habitat types where *P. contorta* is a persistent dominant sometimes maintained by fire, to moist lower subalpine habitat types with *A. lasiocarpa* and *P. engelmannii*, where fires are infrequent but severe with long-lasting effects (Fischer and Bradley, 1987).

The Cooney Ridge wildfire was one of several large wildfire events that occurred during the 2003 fire season in western Montana (Figure 8.1 and Figure 8.2). Lightning ignited the wildfire at several locations on 8 August 2003 (Cooney Ridge

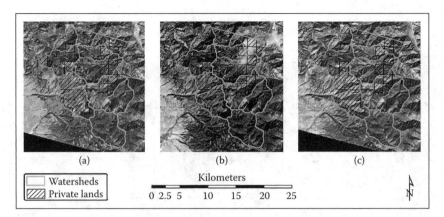

FIGURE 8.2 (See color insert following page 146.) Color infrared composite images of the Cooney Ridge study area (a) 1 year before the wildfire (10 July 2002); (b) during the wildfire (31 August 2003) Note the smoke obscuring the image in the northeastern corner of the burned area; and (c) 1 year after the wildfire (25 September 2004).

Complex Fire Narrative, 2003), and despite intensive suppression efforts, it burned 9600 Ha before it was finally contained on 15 October 2003.

GEOGRAPHIC INFORMATION LAYERS

As noted in Chapter 6, information stored in a GIS can be used to aid forest change analyses by constraining or focusing the change detection efforts, with the goal of extracting more complete and accurate information from spectral data. Four GIS layers proved vital in this case study: watersheds, land ownership, wildfire perimeter, and a forest/nonforest classification. The watershed layer was delineated by applying the TerraFlow (http://www.cs.duke.edu/geo*/terraflow) model to a 30-m digital elevation model obtained from the U.S. Geological Survey National Elevation Dataset. An ownership layer from the state of Montana (http://nris.state.mt.us) indicated lands are 72% public national forest and 28% private industrial timberland. The Cooney Ridge wildfire perimeter originated from the Incident Command GIS team at the Incident Command camp where fire suppression operations were based and is dated 26 August 2003, when the wildfire perimeter had reached its maximum extent. A forest/nonforest map was generated by an image analyst at RSAC based on the six reflectance bands from a 10 July 2002 Landsat ETM+ (Enhanced Thematic Mapper Plus) image (Table 8.1), using commercial *See5* software for thematic classification (www.rulequest.com). The RSAC image analyst trained the classification with 50 forest and 50 nonforest sample points selected across a broader image subset (approximately three times the size) surrounding the study area; classification accuracy was estimated to be 99% (with 1 point misclassified). In this case, *nonforest* was defined as land cover not dominated by green vegetation canopy at the time of image acquisition and as such includes clearcuts, possibly other recent stand-replacing harvest treatments, or natural openings with little vegetation cover, such as rock outcroppings or meadows. The land ownership, burned/unburned, and forest/non-

TABLE 8.1
Landsat-5 (TM) and Landsat-7 (ETM+) Images Used to Characterize 1995–2002 Harvest and 2003 Wildfire Disturbances

Image date	Sensor	Path/row	Indices used	Condition indicated
31 July 1995	TM	41/28	NDVIc	Preharvest
9 September 2001	ETM+	41/27	NBR, NDVIc	Prefire
10 July 2002	ETM+	41/27	NDVIc	Postharvest
31 August 2003	TM	41/27	NBR	Immediate postfire
25 September 2004	TM	41/27	NBR, NDVIc	One year postfire

forest GIS layers were intersected in *ArcInfo*, and the percentage of each within the study area was calculated with an *Excel* pivot table.

IMAGE PREPROCESSING

We acquired five Landsat images (Table 8.1). All five images had been terrain corrected using digital elevation models to correct for relief displacement National Landsat Archive Production (NLAPS) format and were projected to UTM (Zone 11 North). *Imagine* (Leica Geosystems Geospatial Imaging, Norcross, Georgia,) was used to perform all image processing functions.

Calculation of radiance is the fundamental step in standardizing raw image data from multiple sensors to a common radiometric scale (Chander and Markham, 2003). Raw digital number values of spectral bands were converted to radiance values (NASA, 1989). To reduce between-scene variability, spectral radiance was converted to top-of-atmosphere reflectance. This conversion accounted for variable sensor gains and biases, sun angles, earth-sun distances, and solar spectral irradiances (Coops et al., Chapter 2, this volume).

The NDVIc and NBR spectral indices were calculated from the Landsat bands as follows,

$$NDVIc = (B4 - B3)/(B4 + B3) * [1 - (B5 - B5_{min})/(B5_{max} - B5_{min})] \quad (8.1)$$

$$NBR = (B4 - B7)/(B4 + B7) \quad (8.2)$$

where B3 = red band, B4 = near-infrared band, and B5 and B7 are the two Landsat TM (Thematic Mapper) and ETM+ middle-infrared bands. The $B5_{min}$ and $B5_{max}$ constants used to "correct" NDVI (thus calculating NDVIc) are the full-scene minimum and maximum reflectance values in Band 5, respectively, and are assumed to correspond to complete tree canopy closure and openness, respectively.

IMAGE ANALYSES

The calibrated later-date NDVIc and NBR images were subtracted from the calibrated earlier dates to produce dNDVIc and dNBR (delta, d) images. A mask layer was constructed from the 31 July 1995 TM scene edge and the boundaries for 11 watersheds of the same mixed-conifer forest type, which encompassed most of the wildfire perimeter (Figure 8.1). It was necessary to define the study area consistently in this manner, using density slices, to generate comparable mean and standard deviation statistics across all layers for the purpose of threshold-based classifications. Pixels exceeding two standard deviations from the mean clearly indicated the most pronounced land cover change (i.e., stand-replacing harvest or fire) based on visual inspection of the density slice results. All of the distributions were skewed in the direction of the disturbance, and a negligible few to none of the pixels on the opposite sides of the distributions exceeded two standard deviations from the mean, so the output layers were limited to two classes in all cases (i.e., stand-replacing disturbance or not).

The edges of the two output classes after density slicing were heavily pixilated, so an edge-smoothing utility was applied to smooth the class boundaries while also eliminating single-pixel misclassifications. This caused the number of pixels belonging to the minority (disturbance) class to change by an average of 8%. We did not consider this problematic because our intent was not to map the area disturbed accurately, but to define patches consistently and objectively where disturbance effects were most pronounced. The cleaned raster image classes were then converted into vector polygons on which patch metrics could be generated.

PATCH ANALYSES

Many patch metrics are available, although they are often highly intercorrelated (Gustafson, 1998; Riitters et al., 1995). Based on our objective of characterizing landscape pattern effects due to stand-replacing harvest and fire disturbances, a review of quantifying landscape spatial pattern with patch metrics (Gustafson, 1998), and an analysis of landscape pattern change through time across forested landscapes in the region (Hessburg et al., 2000), we selected nine metrics that were thought to be readily interpretable and relevant (Table 8.2). Elkie et al. (1999) provide full details regarding *ArcView Patch Analyst* functions (ESRI, Redlands, CA).

The patch metrics were imported into *R* (R Development Core Team, 2004) for Student *t* tests to test for significant differences in the patch metrics between selected polygon layers of interest. Basing these tests on the entire polygon layers left too few degrees of freedom to produce reliable results. Therefore, more meaningful comparisons were made by partitioning the polygon layers by watershed and treating the watersheds as replicates, which greatly improved the available degrees of freedom to enable robust comparisons.

RESULTS

EXTENT OF FOREST HARVEST AND FIRE DISTURBANCES

We considered the nonforest areas in Figure 8.1 to be predominantly indicative of recent harvest disturbance (some areas such as rocky outcroppings or meadows do not support forest cover). Similarly, we considered the area within the wildfire

TABLE 8.2
Patch Metrics Used to Characterize Size, Edge, and Shape Complexity of Stand-Replacing Harvest and Fire Disturbance Patches

Patch metric	Description
TA	Total area
NP	Number of patches
MNPS	Mean patch size
MDPS	Median patch size
PSSD	Patch size standard deviation
TE	Total edge
MPE	Mean patch edge
AWMSI	Area-weighted mean shape index[a]
MPAR	Mean perimeter:area ratio

[a] AWMSI is a measure of shape complexity. AWMSI equals one when all patches are circular (polygons) or square (grids) and is greater than one when shapes are more complex; individual patch area weighting is applied to each patch. Because larger patches tend to be more complex than smaller patches, area-weighted measures have the effect of determining patch shape complexity independent of patch size (Elkie et al., 1999).

perimeter to have predominantly burned (although some areas did not burn). Because these generalizations should apply equally to both private and public lands, one could assume that land ownership should have no effect on disturbance. Figure 8.3 suggests that private lands were relatively more disturbed than public lands regarding both harvest and fire. Student t tests conducted across the 11 paired watersheds indicated that indeed a significantly higher proportion of private lands (and significantly lower proportion of public lands) was nonforest than would be expected based on the observed nonforest proportion in each watershed without regard to ownership (Table 8.3). However, observed versus expected proportions of private (or public) lands that were inside the wildfire perimeter did not significantly differ (Table 8.3).

PATCH CHARACTERISTICS OF STAND-REPLACING DISTURBANCES

The image differencing and density slicing operations resulted in two NDVIc layers and one dNDVIc layer considered most indicative of stand-replacing harvest prior to 2002 and two NBR layers, two dNBR layers, and one dNDVIc layer considered most indicative of stand-replacing fire from the 2003 wildfire (Figure 8.4 and Figure 8.5). The patch metrics generated on these eight polygon layers quantified patch size, edge, and shape complexity (Table 8.4) of stand-replacing harvest and fire disturbances in this study area over the past decade.

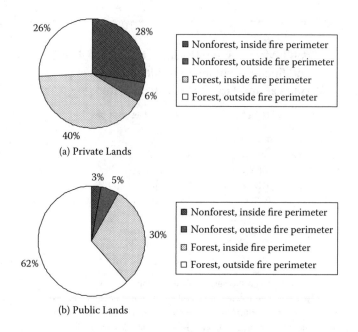

(a) Private Lands

(b) Public Lands

FIGURE 8.3 Observed proportions of (a) private and (b) public lands that were nonforest or forest and inside or outside the wildfire perimeter.

TABLE 8.3
Student *t*-Test Results Comparing Observed versus Expected Proportions of Private and Public Lands that Were Nonforest (and Likely Harvested) or Inside the Wildfire Perimeter (and Likely Burned)

Land category	|*t*| value	*p* value	Significance[a]
Private lands			
Observed versus expected, nonforest	2.8761	.0165	*
Observed versus expected, inside fire perimeter	0.1509	.8830	ns
Public lands			
Observed versus expected, nonforest	2.4039	.0371	*
Observed versus expected, inside fire perimeter	0.5481	.5956	ns

Note: The comparisons were paired across all 11 watersheds.

[a] * = $p < .05$; ns = not significantly different.

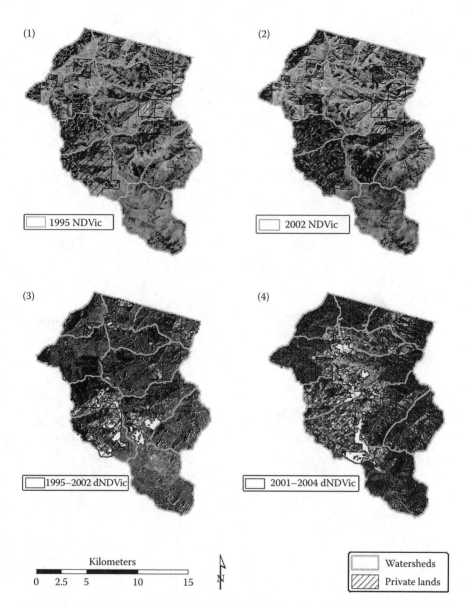

FIGURE 8.4A (See color insert following page 146.) Stand-replacing disturbance maps: (1) 31 July 1995 NDVIc; (2) 10 July 2002 NDVIc; (3) 31 July 1995 to 10 July 2002 dNDVIc; and (4) 9 September 2001 to 25 September 2004 dNDVIc. The NDVIc-derived polygons indicate patches with minimal forest biomass (Maps 1 and 2), and the dNDVIc-derived polygons indicate patches of stand-replacing disturbance before the 2003 wildfire (Map 3) or as a result of the 2003 wildfire (Map 4). The NDVIc-derived patches are more than two standard deviations below the mean image value, while the dNDVIc-derived patches are more than two standard deviations above the mean image value. *Continued.*

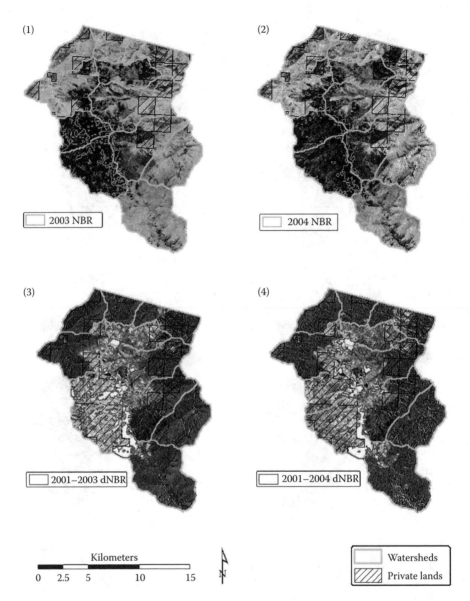

FIGURE 8.4B (See color insert following page 146.) Stand-replacing disturbance maps: (1) 31 August 2003 NBR; (2) 25 September 2004 NBR; (3) 9 September 2001 to 31 August 2003 dNBR; and (4) 9 September 2001 to 25 September 2004 dNBR. The NBR-derived polygons indicate patches with minimal postfire green vegetation cover (Maps 1 and 2), and the dNBR-derived polygons indicate patches of severe fire-induced tree mortality due to the 2003 wildfire (Maps 3 and 4). The NBR-derived patches are more than two standard deviations below the mean image value, while the dNBR-derived patches are more than two standard deviations above the mean image value.

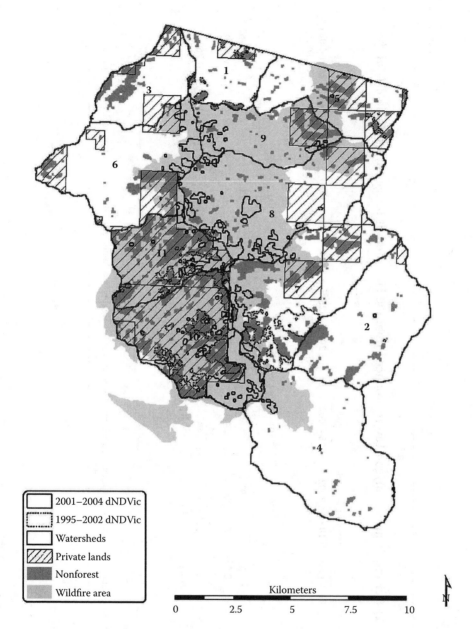

	2001–2004 dNDVIc
	1995–2002 dNDVIc
	Watersheds
	Private lands
	Nonforest
	Wildfire area

Kilometers

0   2.5   5   7.5   10

N

FIGURE 8.5 Juxtaposition of 1995–2002 dNDVIc polygons indicative of stand-replacing disturbance prior to the 2003 wildfire and 2001–2004 dNDVIc polygons indicative of stand-replacing disturbance due to the 2003 wildfire in relation to ownership, the forest/nonforest classification, and the area bounded by the wildfire perimeter (all lands not otherwise labeled are unburned forest on public land). Watersheds are numbered as in Figure 8.1.

TABLE 8.4
Patch Metrics (Table 8.2) Derived from the Polygon Layers Produced from the Five Landsat Image Layers (Table 8.1) of the Entire Study Area

Patch metric	Harvest related			Burn related				
	31 Jul 1995 NDVIc	10 Oct 2002 NDVIc	1995–2002 dNDVIc	31 Aug 2003 NBR	25 Sep 2004 NBR	2001–2003 dNBR	2001–2004 dNBR	2001–2004 dNDVIc
Total area (Ha)	1,098	700	896	842	782	1,209	1,118	1,128
Number of patches	244	178	101	102	53	103	74	93
Mean patch size (Ha)	4.50	3.93	8.87	8.25	14.76	11.74	15.11	12.13
Median patch size (Ha)	1.25	1.62	1.08	1.71	1.71	2.07	2.16	1.62
Patch size standard deviation (Ha)	10.3	8.6	24.4	22.6	49.1	48.1	60.3	48.3
Total edge (m)	222,600	145,440	117,240	134,400	89,100	152,100	118,020	131,760
Mean patch edge (m)	912.3	817.1	1,160.8	1,317.6	1,681.1	1,476.7	1,594.9	1,416.8
Area-weighted mean shape index	1.98	1.74	2.21	2.60	2.95	2.94	2.89	2.84
Mean perimeter area ratio (m/Ha)	504.7	521.8	598.8	404.5	382.3	338.9	358.0	401.1

Note: The three leftmost columns generally indicate stand-replacing harvest patches, while the five rightmost columns generally indicate stand-replacing fire patches.

CONTRASTS BETWEEN STAND-REPLACING HARVEST AND FIRE DISTURBANCE PATCHES

Comparing the patch metrics from delta (difference) images to those from the single-date images (Table 8.1) used to derive the delta images would have violated assumptions of statistical independence, which limited the number of comparisons that could be conducted. The most robust Student t tests compared patch metrics averaged from the two NDVIc layers to patch metrics averaged from the two NBR layers. The NBR patches indicative of stand-replacing fire had significantly higher mean patch size and mean patch edge than the NDVIc patches indicative of stand-replacing harvest (Table 8.5).

Student t tests contrasting the two independent NDVIc layers (pairing all 11 watersheds) revealed fewer patches ($p =.0153$) and less total edge ($p =.0445$) in 2002 than in 1995. This may indicate fewer recent stand-replacing harvest patches in 2002 than in 1995. Comparisons of the independent 2003 and 2004 NBR layers (pairing the 7 watersheds with stand-replacing fire patches) showed no significant differences. Finding more significant differences in the NDVIc layer contrast than in the NBR layer contrast is to be expected given that the NDVIc layers were derived from images 7 years apart, while the NBR layers were derived from images only 1 year apart. Comparisons of the independent 1995–2002 dNDVIc and 2001–2004 dNDVIc layers (pairing the 8 watersheds with patches in both layers) again found no significant differences.

TABLE 8.5
Student t-Test Results Contrasting Patch Metrics Averaged from the 31 July 1995 and 10 July 2002 NDVIc Layers Indicating Stand-Replacing Harvest Patches with Patch Metrics Averaged from the 31 August 2003 and 25 September 2004 NBR Layers Indicating Stand-Replacing Fire Patches

Patch metric	$|t|$ value	p value	Significance[a]
Total area	0.6467	.5417	ns
Number of patches	2.3387	.0580	ns
Mean patch size	4.8254	.0029	**
Median patch size	1.9104	.1047	ns
Patch size standard deviation	1.3474	.2265	ns
Total edge	0.0286	.9781	ns
Mean patch edge	6.2771	.0008	***
Area weighted mean shape index	1.5198	.1794	ns
Mean perimeter area ratio	1.9143	.1041	ns

Note: The tests were paired across the seven watersheds with patches in all four layers.

[a] *** = $p <.001$; ** = $p <.01$; ns = not significantly different.

DISCUSSION

REMOTE SENSING OF FOREST PATTERN AND DISTURBANCE PROCESSES

Satellite images provide a discrete snapshot in time, while landscape disturbance processes are continuous. Multitemporal images (e.g., both pre- and postdisturbance) are generally preferred given an ability to capture disturbance processes because delta images provide a viewable measure of land cover change rather than a snapshot of land cover condition (White et al., 1996). However, care must be taken in image selection for multitemporal analysis for two reasons. First, the image sensors should be compatible. The Landsat data record is most useful given its current length of more than 34 years, which is commensurate with the temporal scale of many forest disturbance processes.

The second reason lies not with the sensor but with the scene. Other vectors of change are captured in delta images besides the disturbance processes of interest. Topographic shadows dramatically affect spatial patterns in rugged terrain such as in our study area, making it highly desirable to choose pre- and postdisturbance images with similar solar illumination conditions. We chose a 9 September 2001 prefire image because it much more closely matched the acquisition months of our two postfire images than the 10 July 2002 image, even though the latter was acquired more recently before the fire. For the same reason, we chose to subtract the 10 July 2002 image from the 31 July 1995 image to indicate prefire stand-replacing disturbance. The months from July to September are typically dry in the northern Rocky Mountains, which greatly influences vegetation phenology. Southern aspects are relatively drier, with sparser tree cover, making the background reflectance more influential and seasonally dynamic. Provided such caveats can be met, delta images are more informative than single-date images for characterizing disturbance.

The 2001–2004 dNBR (Figure 8.4B) and 2001–2004 dNDVIc (Figure 8.5) polygon layers exhibit a highly similar pattern. This is to be expected given that both indices originated from the same source images, and NBR and NDVIc are highly correlated because they share the same near-infrared band. We chose NDVIc over NDVI to indicate forest biomass based not only on literature support (Nemani et al., 1993; Pocewicz et al., 2004) but also because NDVIc has less in common with NBR than NDVI (compare Equation 8.1 and Equation 8.2). While BAER teams prefer NBR over NDVI for the greater sensitivity of NBR to soil effects (Parsons, 2003), both indices are highly sensitive to vegetation cover (Hudak, Morgan, et al., 2004; Hudak, Robichaud, et al., 2004). At the Cooney Ridge wildfire, NBR and dNBR-based BARC maps used by BAER teams showed the largest proportion of high burn severity along the ridge forming the eastern boundary of Watersheds 10 and 11, which our NBR and dNBR layers corroborate (Figure 8.4B).

PATCH CHARACTERISTICS OF STAND-REPLACING HARVEST AND FIRE DISTURBANCE

We chose a consistent and objective threshold of two standard deviations from the mean to define stand-replacing disturbance, whether induced by harvest or fire. We felt that the absence of spatially explicit field data to indicate subtler timber harvest practices (e.g., thinning) or less-severe burns justified choosing a conservative threshold for defining stand-replacing disturbance. The *location* of the major stand-replacing harvest and fire disturbance patches, as indicated in Figure 8.5, matched our observations on the ground. However, we believe that the *extent* of the mapped stand-replacing disturbance patches is conservative (Figure 8.5) based on our field observations of postfire effects both immediately and one year after the fire; they are very conservative compared to the extent of harvest and fire disturbances suggested by the geographic layers (Figure 8.1), which actually much more closely resemble our impressions on the ground.

Like virtually all large wildfires, the Cooney Ridge postfire landscape was very heterogeneous, with patches varying in size and shape. Many patches within the wildfire perimeter were lightly burned, and some remained unburned. We do not recommend our image analysis approach of classifying pronounced departures from the mean for accurate mapping of burn area extent (see Hudak and Brockett, 2004), which becomes difficult at low severities. Similarly, this approach is not ideal for mapping the extent of timber harvest areas (see Cohen et al., 1998) as many partial cuts will be omitted. Encouragingly, the 1995 and 2002 NDVIc polygon layers that we considered indicative of stand-replacing harvest (Figure 8.4A) show a pattern closely matching (but with more limited extent) that of nonforested lands (which would include more partial cutting) mapped by the RSAC image analyst using *See5* thematic classification software (Figure 8.5).

Results from this case study demonstrated that stand-replacing harvest patches, on average, had significantly less area (mean patch size) and edge (mean patch edge) than stand-replacing fire patches (Table 8.4 and Table 8.5). In general, clearcutting results in forest pattern characterized by smaller patch sizes, smaller patch perimeter lengths, greater distances between patches, more edge habitat, and less interior habitat (Reed et al., 1996) when compared to patterns created by natural processes such as fire, insect outbreak, avalanches, and blowdowns (Tinker and Baker, 2000).

Stand-replacing harvest (Cohen et al., 1998; Healey et al., Chapter 3, this volume) and fire disturbances (Hessburg et al., 2000) may be the principal current determinants of landscape pattern. Prior to European settlement and significant timber-harvesting activity, fire was the principal disturbance shaping landscape pattern. Undoubtedly, topography and other disturbances such as insects, disease, and wind also influenced forest pattern, yet fire effects are coupled to all of these. Timber harvest, fire, and roads are now the principal determinants of landscape pattern on many private and public lands, particularly in mid- to high-elevation mixed-conifer forests that have many roads to facilitate fire detection and suppression (Hessburg et al., 2000; Linke et al., Chapter 1, this volume), which are high priorities in landscapes subject to logging and recreational use. Moreover, the primary spatial scale of structural variation in forests today is at the stand level due to harvesting "footprints," while historically the primary scale of forest structural variation may have been broader and closer to the scale at which burn patches vary across the landscape.

INTERACTION OF FOREST HARVEST AND FIRE DISTURBANCE PROCESSES

Our intent was to consistently and objectively define patches resulting from forest harvest or fire disturbances. While consistency and objectivity are always advisable, they are especially important when presenting results without the benefit of geolo-cated validation data for accuracy assessment. However, we have been heavily involved with wildfire research at Cooney Ridge, where we have measured prefire fuels and active fire characteristics at one site and made extensive postfire effects measurements at this and several other sites distributed across the entire landscape (Morgan et al., 2004). These field data were gathered prior to this analysis and to meet different objectives, but in the process of crisscrossing the area while conduct-ing fieldwork, we became very familiar with the entire Cooney Ridge postfire landscape. The significantly higher association (Figure 8.3, Table 8.3) of private lands with largely harvested lands, compared to public lands, was confirmed by our observations on the ground, the patterns visually apparent in the satellite imagery (Figure 8.2), and the consistent and objective density slicing approach we used to delineate stand-replacing disturbance events (Figure 8.4 and Figure 8.5).

The most unexpected result from the patch metrics analysis is the similarity between the 1995–2002 dNDVIc and 2001–2004 dNDVIc patch metrics. No signif-icant differences were found across all nine metrics. The 1995–2002 dNDVIc map shows that the areas of greatest vegetation change within the study area in this time interval occurred in Watersheds 7, 10, and 11. The large polygons in Watershed 7 can be attributed to the 700-Ha 1998 Gilbert Creek 2 fire (Gilbert Creek Fire Incident Action Plan, 1998), part of which reburned through the 226-Ha 1985 Gilbert Creek 1 fire. Both fires occurred in early September (Ed Mathews, U.S. Forest Service Missoula Fire Sciences Lab, email, 1 December 2005). The large polygons in Watersheds 10 and 11 can be attributed to large clearcuts on the private industrial forest land that comprises most of these watersheds (Figure 8.5). Enough time elapsed since these disturbances to allow shrubs and herbaceous vegetation to recover, thus increasing the 10 July 2002 NDVIc values sufficiently to escape detection in the single-date density slice of this image (Rogan et al., 2002). This exemplifies the value of delta images over single-date images for disturbance map-ping, especially as time elapses until the acquisition of the postdisturbance image (Hudak and Brockett, 2004).

Timing of image acquisitions heavily influenced our results and interpretation. The clearcut areas in Watersheds 10 and 11 were mapped as severely burned using NBR derived from immediate postfire imagery that RSAC used to produce a BARC map (Stone et al., 2004). Many of these same clearcuts were no longer mapped as severely burned when NBR and dNBR were derived from postfire imagery acquired one year later (Figure 8.4B and Figure 8.5). This exemplifies the merit of one-year postfire images for extended assessments of burn severity because the degree of postfire vegetation regrowth is in itself a very useful indicator of ecological impact (Key and Benson, in press).

Most of the areas mapped as severely burned in our analysis were on steep, upper slopes adjacent to and above clearcuts. The 1995–2002 dNDVIc and 2001–2004 dNDVIc polygons clearly do not overlap (Figure 8.5) because following a clearcut there is little biomass remaining to burn compared to a mature forest. However, what is more remarkable is the obvious *adjacency* of the polygons in these two independently derived layers. The adjacency of the 1995–2002 dNDVIc polygons on the eastern side of Watersheds 10 and 11 to the 2001–2004 dNDVIc polygons immediately east (i.e., on either side of the ownership boundary) matches our field observations. Strong westerly winds on 16 and 17 August 2003 caused extreme fire behavior and the fastest fire progression of all days on the Cooney Ridge wildfire based on unpublished GIS data obtained from fire managers (Stone et al., 2004). We believe that availability of abundant dry fuel stemming from recent clearcuts on the private lands in Watersheds 10 and 11 coupled with extremely low fuel moistures, local topography, and very hot, dry, windy weather all contributed to the rapid advance of intensely burning fire from the clearcut private lands into the standing timber on public land. This resulted in the large, severely burned patches along the ridge defining the eastern edge of Watersheds 10 and 11 (Figure 8.5). In many mid- to high-elevation forests common in the northern Rocky Mountains, weather and topography rather than fuels are often the primary variables determining fire size and severity (Bessie and Johnson, 1995; Sherriff et al., 2001; Turner et al., 1994).

Together, the large clearcuts on private lands and the extensively burned areas on both private and public lands created large, relatively homogeneous patches with few trees. Although shrubs and grasses will rapidly regrow, the lack of tree cover, especially on steep slopes, could contribute to soil erosion. Postlogging tree planting and postburn rehabilitation are designed to hasten tree establishment and to mitigate possible soil erosion. To the credit of local managers, we did observe many newly planted tree seedlings in Watershed 11 one year after the wildfire.

CONCLUSIONS

This case study illustrates the importance of landscape context in determining severe burn patterns. Fuels, weather, and topography interact to determine active fire behavior and subsequent postfire effects (Pyne et al., 1996). Unfortunately, current understanding of these interactions is limited. Land use features such as roads and clearcuts can fragment forested landscapes (Bresee et al., 2004). Fire management decisions also alter landscape pattern. Fire managers are very successful at suppressing the vast majority of fires, so most are small. Hessburg et al. (2000) quantified a high degree of change and variability in forest landscape pattern over 60 years across Idaho, Montana, and Washington and attributed this to the combined effects of fire exclusion and other land uses. In further analysis of their data, Black et al. (2003) found that changes in forest patterns across mountainous landscapes were correlated with both human and biophysical factors.

Fire and other disturbances have played important ecological roles in western coniferous forest ecosystems. In extreme years, especially after prolonged drought, extensive areas burn across the western United States (Swetnam and Betancourt, 1990, 1998). Such years account for the majority of the area burned (Strauss et al., 1989) and the greatest threats to people and property (Maciliwain, 1994). Fuel management through logging or other means will be less effective when drought and weather conditions are extreme, as they were in western Montana in 2003. One of the clearest lessons from history is that fires have always occurred, and they will continue to occur despite our efforts to detect and suppress them (Morgan et al., 2003). In most forest ecosystems in western North America, biomass production exceeds decomposition; this accumulated biomass fuels fires when lightning or people ignite fires in hot, dry, windy conditions. An understanding of where fires are more likely to be severe would help to strategically locate and design fuel management treatments where they will be most effective. Such an understanding would also be helpful in strategic fire suppression, fire mitigation, and postfire rehabilitation decisions.

Like all real landscapes, the Cooney Ridge landscape is unique. Thus, it would be misguided to generalize our case study results and interpretation to other landscapes, which have their own unique contexts. Yet, the disturbance processes observed at Cooney Ridge are common to other forested landscapes shaped by timber harvest and fire, as nearly all forested landscapes are to some degree. The recent, dramatic disturbance history at Cooney Ridge creates a fertile setting for exploring how human and natural disturbances interact to shape landscape pattern. This case study may raise more questions than it answers; in fact, we hope that it does. We encourage others to think about how they might also use remote sensing and GIS tools for quantifying landscape patterns, which can provide a window for better understanding of landscape disturbance processes.

ACKNOWLEDGMENTS

This research was supported in part by funds provided by the Rocky Mountain Research Station, Forest Service, U.S. Department of Agriculture, in cooperation with the University of Idaho (JVA 03-JV-11222065–279), through funding from the USDA/USDI Joint Fire Science Program, Project 03-2-1-02, entitled, "Assessing the Causes, Consequences and Spatial Variability of Burn Severity: A Rapid Response Project." We thank Jeffrey Evans at the RMRS Moscow Forestry Sciences Laboratory for delineating the watershed boundaries; Bonnie Ruefenacht at RSAC for producing the *See5* forest/nonforest classification; Ed Mathews and Colin Hardy at the RMRS Missoula Fire Sciences Laboratory for identifying the Gilbert Creek fires; Rudy King for statistical advice; and Mike Wulder, Dennis Ferguson, Henry Shovic, and three anonymous reviewers for their helpful comments.

REFERENCES

Bessie, W.C. and Johnson, E.A. (1995). The relative importance of fuels and weather on fire behavior in subalpine forests. *Ecology, 76*, 747–762.

Black, A.E., Morgan, P., and Hessburg, P.F. (2003). Social and biophysical correlates of change in the landscape structure of forests in the interior Columbia River Basin. *Ecological Applications, 13*, 57–67.

Bresee, M.K., Le Moine, J., Mather, S., Brosofske, K.D., Chen, J., Crow, T.R., and Rademacher, J. (2004). Disturbance and landscape dynamics in the Chequamegon National Forest, Wisconsin, USA, from 1972 to 2001. *Landscape Ecology, 19*, 291–309.

Brockett, B.H., Biggs, H.C., and van Wilgen, B.W. (2001). A patch mosaic burning system for conservation areas in southern African savannas. *International Journal of Wildland Fire, 10*, 169–183.

Chander, G. and Markham, B. (2003). Revised Landsat-5 TM radiometric calibration procedures and postcalibration dynamic ranges. *IEEE Transactions on Geoscience and Remote Sensing, 41*, 2674–2677.

Cohen, W.B., Fiorella, M., Gray, J., Helmer, E., and Anderson, K. (1998). An efficient and accurate method for mapping forest clearcuts in the Pacific Northwest using Landsat imagery. *Photogrammetric Engineering and Remote Sensing, 64*, 293–300.

Cooney Ridge Complex Fire Narrative. (2003). MT-SWS-000149. USDA Forest Service Region 6. 22p.

De Bano, L.F., Neary, D.G., and Ffolliott, P.F. (1998). *Fire's Effects on Ecosystems*. John Wiley and Sons, New York. 333p.

Elkie, P., Rempel, R. and Carr, A. (1999). *Patch Analyst User's Manual*. Ontario Ministry of Natural Resources. Northwest Science and Technology, Thunder Bay, ON, Canada. TM-002. 16pp plus Appendix.

Fischer, W.C. and Bradley, A.F. (1987). *Fire Ecology of Western Montana Forest Habitat Types*. USDA Intermountain Research Station, Ogden, UT. General Technical Report INT-223. 95p.

Gilbert Creek Fire Incident Action Plan. (4 September 1998). Lolo National Forest, MT. 19p.

Graham, R.T. (Tech. Ed.). (2003). *Hayman Fire Case Study*. USDA Forest Service, Ogden, UT. GTR-RMRS-114. 396p.

Gustafson, E.J. (1998). Quantifying landscape spatial pattern: what is the state of the art? *Ecosystems, 1*, 143–156.

Hessburg, P.F., Smith, B.G., Salter, R.B., Ottmar, R.D., and Alvarado, E. (2000). Recent changes (1930s–1990s) in spatial patterns of interior northwest forests, USA. *Forest Ecology and Management, 136*, 53–83.

Hudak, A.T. and Brockett, B.H. (2004). Mapping fire scars in a southern African savannah using Landsat imagery. *International Journal of Remote Sensing, 25*, 3231–3243.

Hudak, A.T., Fairbanks, D.H.K., and Brockett, B.H. (2004). Trends in fire patterns in a southern African savanna under alternative land use practices. *Agriculture, Ecosystems and Environment, 101*, 307–325.

Hudak, A., Morgan, P., Stone, C., Robichaud, P., Jain, T., and Clark, J. (2004). The relationship of field burn severity measures to satellite-derived burned area reflectance classification (BARC) maps. *American Society for Photogrammetry and Remote Sensing Annual Conference Proceedings* (pp. 96–104). CD-ROM. ASPRS, Bethesda, MD.

Hudak, A., Robichaud, P, Evans, J., Clark, J., Lannom, K., Morgan, P., and Stone, C. (2004). Field validation of burned area reflectance classification (BARC) products for post fire assessment. *Proceedings of the Tenth Biennial Forest Service Remote Sensing Applications Conference*. CD-ROM. 13p.

Johnson, E.A., Morin, H., Miyanishi, K., Gagnon, R., and Greene, D.F. (2003). A process approach to understanding disturbance and forest dynamics for sustainable forestry. In P.J. Burton, C. Messier, D.W. Smith, and W.L. Adamowicz (Eds.), *Towards Sustainable Management of the Boreal Forest*. (pp. 261–306). National Research Council Research Press, Ottawa, ON, Canada. 1039p.

Key, C.H. and Benson, N.C. (In press). Landscape assessment: ground measure of severity, the composite burn index; and remote sensing of severity, the normalized burn ratio. In D.C. Lutes, R.E. Keane, J.F. Caratti, C.H. Key, N.C. Benson, and L.J. Gangi (Eds.), *FIREMON: Fire Effects Monitoring and Inventory System*. USDA Forest Service, Rocky Mountain Research Station, Fort Collins. General Technical Report.

Kimmins, J.P. (1997). *Forest Ecology: A Foundation For Sustainable Management*. (2nd ed.). Prentice-Hall, Englewood Cliffs, NJ. 596p.

Lewis, S.A., Wu, J.Q., and Robichaud, P.R. (2006). Assessing burn severity and comparing soil water repellency, Hayman Fire, Colorado. *Hydrological Processes, 20,* 1–16.

Maciliwain, C. (1994). Western inferno provokes a lot of finger-pointing but not much action. *Science, 370,* 585.

Mladenoff, D.J., White, M.A., Pastor, J., and Crow, T.R. (1993). Comparing spatial pattern in unaltered old-growth and disturbed forest landscapes. *Ecological Applications, 3,* 294–306.

Moore, M.M., Covington, W.W., and Fule, P.Z. (1999). Reference conditions and ecological restoration: a southwestern ponderosa pine perspective. *Ecological Applications, 9,* 1266–1277.

Moreno, J.M. and Oechel, W.C. (1989). A simple method for estimating fire intensity after a burn in California chaparral. *Ecologica Plantarum, 10,* 57–68.

Morgan, P., Defosse, G.E., and Rodriguez, N.F. (2003). Management implications of fire and climate changes in the western Americas. In T.T. Veblen, W.L. Baker, G. Montenegro, and T.W. Swetnam (Eds.), *Fire and Climatic Change in Temperate Ecosystems of the Western Americas* (pp. 413–440). Springer-Verlag, New York. Ecological Studies 160.

Morgan, P., Hudak, A., Robichaud, P., and Ryan, K. (2004). *Assessing the Causes, Consequences and Spatial Variability of Burn Severity: A Rapid Response Proposal.* Retrieved from http://www.cnrhome.uidaho.edu/burnseverity. JFSP project 03–2-1–02. Progress Report. September 2004.

NASA. (1989). *Landsat-7 Science Data User's Handbook.* NASA/Goddard Space Flight Center, Greenbelt, MD.

Nemani, R.R., Pierce, L., Running, S., and Band, L. (1993). Forest ecosystem processes at the watershed scale: sensitivity to remotely-sensed leaf area index estimates. *International Journal of Remote Sensing, 14,* 2519–2534.

Parsons, A. (2003). *Burned Area Emergency Rehabilitation (BAER) Soil Burn Severity Definitions and Mapping Guidelines.* Burn Severity Definitions/Guidelines Draft. 12p. USDA Forest Service, Remote Sensing Applications Center. Salt Lake City, UT. (Unpublished report.)

Pocewicz, A.L., Gessler, P., and Robinson, A. (2004). The relationship between effective plant area index and Landsat spectral response across elevation, solar insolation, and spatial scales in a northern Idaho forest. *Canadian Journal of Forest Research, 34,* 465–480.

Pritchett, W.L. and Fisher, R.F. (1987). *Properties and Management of Forest Soils.* 2nd ed. John Wiley and Sons, New York. 494p.

Pyne, S.J., Andrews, P.L., and Laven, R.D. (1996). *Introduction to Wildland Fire.* 2nd ed. John Wiley and Sons, New York. 769p.

R Development Core Team. (2004). *R: A Language and Environment for Statistical Computing.* R Foundation for Statistical Computing, Vienna, Austria. Retrieved from http://www.R-project.org. October 24, 2004.

Reed, R.A., Johnson-Barnard, J., and Baker, W.L. (1996). Contribution of roads to forest fragmentation in the Rocky Mountains. *Conservation Biology, 10,* 1098–1106.

Riitters, K.H., O'Neill, R.V., Hunsaker, C.T., Wickham, J.D., Yankee, D.H., Timmins, S.P., Jones, K.B., and Jackson, B.L. (1995). A factor analysis of landscape pattern and structure metrics. *Landscape Ecology, 10,* 23–39.

Rogan, J., Franklin, J., and Roberts, D.A. (2002). A comparison of methods for monitoring multitemporal vegetation change using Thematic Mapper imagery. *Remote Sensing of Environment, 80,* 143–156.

Rowe, J.S. (1983). Concepts of fire effects on plant individuals and species. In R.W. DeBano and D.A. MacLeans (Eds.), *The Role of Fire in Northern Circumpolar Ecosystems* (pp. 135–154). John Wiley and Sons, New York. 322p.

Ryan, K.C. (2002). Dynamic interactions between forest structure and fire behavior in boreal ecosystems. *Silva Fennica, 36,* 13–39.

Ryan, K.C. and Noste, N.V. (1985). Evaluating prescribed fires. In J.E. Lotan, B.M. Kilgore, W.C. Fisher, and R.W. Mutch (coordinators), *Proceedings, Symposium and Workshop on Wilderness Fire* (pp. 230–238). November 15–18, 1983, Missoula, MT. USDA Forest Service, General Technical Report INT-182.

Savage, M. and Nystrom Mast, J. (2005). How resilient are southwestern ponderosa pine forests after crown fires? *Canadian Journal of Forest Research, 35,* 967–977.

Schimmel, J. and Granstrom, A. (1996). Fire severity and vegetation response in the boreal Swedish forest. *Ecology, 77,* 1436–1450.

Sherriff, R.L., Veblen, T.T., and Sibold, J.S. (2001). Fire history in high elevation subalpine forests in the Colorado Front Range. *Ecoscience, 8,* 369–380.

Smith, D.M., Larson, B.C., Kelty, M.J., and Ashton, P.M.S. (1996). *The Practice of Silviculture: Applied Forest Ecology.* 9th ed. John Wiley and Sons, New York. 537p.

Stone, C., Hudak, A., and Morgan, P. (2004). Forest harvest can increase subsequent forest fire severity. *International Symposium on Fire Economics, Policy and Planning: A Global Vision* (pp. 154–163). CD-ROM. University of Córdoba, Córdoba, Spain. Editor: Armando González-Cabán.

Strauss, D., Bednar, L., and Mees, R. (1989). Do one percent of the forest fires cause 99% of the damage? *Forest Science, 35,* 319–328.

Swetnam, T.W. and Betancourt, J.L. (1990). Fire–southern oscillation relations in the southwestern United States. *Science, 249,* 1017–1020.

Swetnam, T.W. and Betancourt, J.L. (1998). Mesoscale disturbance and ecological response to decadal-scale climate variability in the American Southwest. *Journal of Climate, 11,* 3128–3147.

Tinker, D.B. and Baker, W.L. (2000). Using the LANDLOG model to analyze the fragmentation of a Wyoming forest by a century of clear-cutting. In R.L. Knight, F. W. Smith, S.W. Buskirk, W.H. Romme, and W.L. Baker (Eds.), *Forest Fragmentation in the Southern Rocky Mountains* (pp. 337–358). University Press of Colorado, Boulder, Colorado, USA.

Turner, M.G., Hargrove, W.W., Gardner, R.H., and Romme, W.H. (1994). Effects of fire on landscape heterogeneity in Yellowstone National Park, Wyoming. *Journal of Vegetation Science, 5,* 731–742.

Turner, M.G., Romme, W.H., Gardner, R.H., and Hargrove, W.W. (1997). Effects of fire size and pattern on early succession in Yellowstone National Park. *Ecological Monographs, 67,* 411–433.

White, J.D., Ryan, K.C., Key, C.C., and Running, S.W. (1996). Remote sensing of forest fire severity and vegetation recovery. *International Journal of Wildland Fire, 6,* 125–136.

9 Conclusion: Understanding Forest Disturbance and Spatial Pattern, Information Needs, and New Approaches

Michael A. Wulder and Steven E. Franklin

CONTENTS

CONTEXT, CONVERGENCE, AND INTEGRATION

Remotely sensed and geographical information system (GIS) data may be processed or combined to produce information depicting landscape patterns or structure, including land cover, biophysical or biochemical status of vegetation, or other attributes of interest (e.g., soil moisture). Information regarding the land cover representing a known time period may be produced through image classification. Change detection methods may then be applied to quantify the dynamics that have occurred or are presently active. Pattern analysis — in the form of landscape metrics or landscape pattern indices (LPIs) — may then be applied to the land cover information produced through image classification, the disturbance information produced through change detection approaches, or both. This integration of forest change information with the tools of pattern analysis provides for unique insights into the outcomes of management decisions or disturbance events. Comparisons over space and time are then possible between the patterns that emerge from differing dynamics or against theoretical base conditions. It is this comparison between the patterns that emerge from different disturbances (i.e., fire versus harvesting) that empowers resource managers to consider anthropogenic change within the greater context of disturbance processes active on the landscape.

The disturbance patterns evident through measurement provide insights and encourage inferences to be made regarding the processes operating on the landscape. There is a complex interaction of feedbacks between pattern and processes on the landscape (Linke et al., Chapter 1). When attempting to characterize these feedbacks, an understanding of forest structure and dynamics, appropriate data sources, methods to capture forest disturbances, and subsequent pattern analysis is required. Typically, each of these factors is considered in isolation; however, the interactions between each of these elements create a complex web of considerations that can have a significant impact on the nature and utility of the resulting information. Through an improved understanding of these interactions, it is our view that the near- and long-term prospects for fully integrating forest dynamics and patterns with current forest conditions are encouraging. Our intention in producing this volume was to tie these concepts together using domain-specific clarification (e.g., case studies) and reviews of the key concepts associated with the integration of remote sensing, GIS, and landscape ecology from a forest-monitoring perspective. This integration is beyond methods but points to a problem-solving philosophy — or approach — to arrive at the desired information through consideration and understanding of the process(es) of interest, clear definition of the information needs required to make appropriate management decisions regarding the processes and patterns that result, and then to implement the appropriate remote sensing and GIS data selection and analytical methods to obtain optimal desired or specified output products.

The objective of this concluding chapter is to summarize and synthesize the essence of preceding chapters to provide sufficient context for presentation of some possible future directions envisioned. These future directions will focus on the significant opportunities that will be afforded through the accurate and reliable (perhaps even routine) capture of forest disturbance, in essence, through remote sensing and GIS approaches coupled with emerging spatial pattern analysis techniques.

UNDERSTANDING FOREST LANDSCAPES: REMOTE SENSING AND GIS APPROACHES

Forest succession and growth are well-established concepts, with a theoretical basis and well-defined means for characterization in the field and, increasingly, through remote sensing. From a baseline of information, such as a forest inventory, image classification, or permanent sample plot database, forest growth and succession may be modeled to produce management information products. Forest disturbance, on the other hand, typically must be captured by some independent means; one of the goals of this independent assessment (for example, through sketch mapping of defoliation) is to ensure the quality of information products modeled from a baseline of information. Landscape ecological principles enable an informed view of spatial and temporal patterns by relating pattern and processes and understanding the primary and feedback links.

Linke et al. (Chapter 1) presented some fundamental concepts useful for understanding forest succession, structure, function, and change at the landscape level. The authors provide an overview of the ecological foundations of landscape ecology, a discipline with prominence that has increased in recent years largely as a result of advances in the fields of remote sensing and GIS. This overview is a critical context for the importance of detecting and monitoring changes in forest landscape patterns. Key to this first chapter is an elucidation of the complex reciprocal relationships between forest patterns and processes. The links between the nature of the dynamics occurring in the forest to large-scale processes, such as biodiversity and species composition, are only now being fully realized and understood. For example, it is now widely acknowledged that the conservation of biodiversity requires the collection of new information to complement traditional sources, such as forest inventory, modeling, and field observations. Remotely sensed data, and the storage and processing opportunities resulting from advances in computer technology, offer those interested in biodiversity conservation a suite of unique approaches that, in combination with models, will address the increasing information needs to ensure appropriate management decision making. The integration of a variety of methods and techniques from remote sensing, GIS, and modeling represents a powerful new approach to combining data and knowledge synergistically, utilizing individual geospatial tools, to produce otherwise unavailable and even unsuspected information from the vast and growing databases under assembly for forest management purposes. The knowledge gleaned at specific points and over generations of detailed forest observation can be integrated using these new approaches to produce spatial, mapped outputs. Landscape patterns emerging from differing disturbance regimes, largely as captured with remotely sensed data, can be quantified, compared, and considered under a range of contexts. The ability to consider the wider ecological meaning of emerging spatial patterns and how management or conservation practices can be adjusted to accommodate these new insights are increasingly enabled through this new approach, which relies on the application of specific geospatial technologies.

The basic elements for consideration of the ecology of landscapes are the landscape structure, function, and change (Forman, 1995). Disturbance regimes must be characterized by magnitude, timing, and spatial distribution; each of these elements will have an impact on the stand- (or patch-) and landscape-level character of the forest ecosystem; of course, each of these elements varies by disturbance regime. This variability will create enormous complications when selecting the appropriate data source to capture the change, the timing of data collection, and the grain and extent of the data. This complex interplay of data sources and application needs are described by Coops et al. in Chapter 2; they advise those considering adoption of the remote sensing and GIS approach to forest disturbance and landscape monitoring to know what processes or patterns are of interest and to ensure that the appropriate data and processing methods are employed. *Information needs* are the key driver of both the data and processing methods selection; in other words, first obtain an understanding of the information required; and from this understanding will follow specific constraints for data and methods selection. Iteration of the choices available (from data and methods perspectives) to reach the information goals can then be undertaken considering the various trade-offs (including spatial resolution, image extent, timing, cost). Accommodation can be made to ensure the optimal data and processing approaches are determined for the given application.

Methodological approaches — including the remote sensing and GIS approaches described here — are bounded by information desired in the application and, of course, the data available. For example, while the Landsat sensors have been the most commonly processed data for undertaking spatial pattern analysis (Gergel, Chapter 7), higher spatial resolution sensors now commonly provide data commercially with extents of approximately 100 km^2 composed of 1×1 m pixels. Coops et al. (Chapter 2) illustrate that data are available from increasing numbers of sensors, with a greater variety of spatial and spectral configurations. In addition, the greater number of sensors provides for an improved ability to acquire data over the temporal period of interest.

There is a multitude of methods and specific techniques or algorithms that can be adopted when using remotely sensed and GIS data for the detection and monitoring of forest change. The approach — in general — is based on methods and techniques that are typically designed to reveal spectral or biophysical changes over time. Regardless of the image-processing methods selected, GIS data are increasingly considered essential to aid and augment the methods and to extract accurate disturbance information (Rogan and Miller, Chapter 6). Coops et al. (Chapter 2, Figure 2.5) present a theoretical representation of the increase in accuracy and decrease in confidence intervals associated with the ability to detect differing forest disturbances. As disturbances on the forest landscape become more severe (e.g., increase in size), contiguous detection is more likely (i.e., stand replacing). In contrast, disturbances that are small and heterogeneous over the landscape (i.e., defoliation or partial harvesting) are generally more difficult to detect with remotely sensed data. The ability with which disturbances of various sizes are successfully detected and mapped is of course dependent on the data source selected for the application. Furthermore, the spectral variability associated with non-stand-replacing disturbances is typically greater than that associated with stand-replacing disturbance, making repeat detection of these disturbances less probable and often resulting in lowered precision of these change estimates. As expected, larger and more spatially contiguous disturbances are generally mapped with greater consistency and greater accuracy using a given data source.

Healey et al. (Chapter 3) present an established spectral change approach that has been used successfully in the U.S. Pacific Northwest. The requirement to understand information needs and change types present on the landscape is a key component of their application, and thus they exemplify the remote sensing and GIS approach. The authors suggest that harvest types vary (clearcut, partial, thinning, etc.), with each having a differing impact on the landscape, and each subsequently impacting the resultant spectral properties of a stand. This understanding leads to an incorporation of these silvicultural distinctions in their change maps and help form an understanding of the spatial, temporal, and spectral resolution issues surrounding the remote detection of all types of silvics and harvest disturbance. The notion that different measures of forest removal have the ability to emphasize different structural or silvicultural elements of harvest is presented. As a result, decisions on how to label and differentiate harvest types based on monitoring needs must be integrated into the design of the overall change detection approach.

To complement the spectral change approaches, which are largely based on land cover or forest canopy change, employed in Chapters 3, 5, and 8, Hall et al. in Chapter 4 present a detection and mapping approach based on changes in the biophysical attribute of leaf area index that result from defoliation caused during an insect infestation. A primary goal of insect defoliation mapping is consistency in output products. Also important is an ability to identify a range of defoliation — or possibly damage — classes, usually indicative of severity of defoliation. Spectral change approaches, while successful in characterizing defoliation, do not typically enable damage severity categorization; this is likely a result of high variability — or a lack of consistency — in the relationship between spectral change and foliar change. Similar to the difficulties in capturing subtle changes through partial harvesting approaches, defoliation does not result in a stand replacement and can be difficult to detect. Remote sensing scientists have usually recommended the use of more customized methods rather than off-the-shelf commercially available image-processing tools. Hall et al. (Chapter 4), recommend the leaf area index-based biophysical approach, augmented with field observations and aerial sketch-mapping information, and they show that the remote detection and categorization of aspen defoliation is accurate and reliable. Similar to how differing silvicultural regimes result in a range of landscape conditions, the effect on forest canopies and structure often vary by insect. Hall et al. describe a range of insects and the related damage characteristics and then tie these conditions back to the expected damage characteristics, which enables appropriate data and methods selection. Again, these authors exemplify the link between understanding the disturbance process, the information needs of resource managers, and remote sensing and GIS approaches.

The maturity and consistency of operational change detection activities is exemplified in Clark and Bobbe in Chapter 5. Satellite-based fire mapping is a key component of suppression and mitigation activities of the U.S. Forest Service. These authors show that the outputs are sufficiently useful to justify the considerable investment in infrastructure and personnel. The output products are also consistent in that the users are able to systematically consider and act on the mapped fire locations and severities.

In this book, remotely sensed and GIS data are complementary; GIS data are considered to be the nonspectral digital data included and used in forest mapping and monitoring applications. GIS data are typically described as "collateral" and "ancillary" in the literature, as compared to "primary" data collected with remote sensors. Yet, as clearly shown in this book and in numerous applications almost since the advent of the field of remote sensing, the strength of remotely sensed and GIS data can only be obtained through the complementary nature of these two data sources, and this synergy can best be exploited through integration. Rogan and Miller in Chapter 6 indicate that the integration of remotely sensed and GIS data takes four primary forms: (a) GISs can store multiple data types; (b) GIS analysis and processing enable raster data manipulation and analysis (e.g., buffer/distance operations); (c) remotely sensed data can be manipulated to derive GIS data; and (d) GIS data can be used to guide image analysis to extract more complete and accurate information from spectral data. The focus of Chapter 6 is primarily on the fourth topic, with the support of the previous three methods. The authors emphasize the notion that GIS data and systems can be utilized to produce improved disturbance information from remotely sensed data. Combined advances in computing power, inexpensive disk storage, and software have enabled the increased integration of remotely sensed and GIS data. Rogan and Miller also highlight current research themes that will further the integration of remote sensing and GIS technologies, including expert systems, object-based analyses, simultaneous consideration of spatial and temporal components of change, continuous fields, and persistent diligence regarding the accuracy of information products. Advances in these areas are thought to have potential for generating disturbance information that is in turn more suitable for the increasingly reliable and predictable approaches emerging from pattern analyses.

Gergel (Chapter 7) reviews the expanding discipline of landscape ecology and the related development of LPIs and posits that the discipline of remote sensing and related advances and techniques in other fields are directly responsible for a new ability to ask questions about pattern at broad spatial scales. It is this ability to consider questions representative of broader scales on large landscapes, in contrast to plot-focused traditional ecology and field sampling, which have advanced the discipline of landscape ecology. The understanding and communication of pattern analysis approaches and outcomes are expanding rapidly, with these approaches possibly beneficial to a wide range of environmental science disciplines. To aid in the robust application of pattern analysis tools, Gergel recommends a systematic approach to implementation: (a) clearly state questions and hypotheses, (b) use only nonredundant metrics, (c) examine the frequency distributions of LPIs, (d) elucidate on possible errors or biases, and (e) consider the accuracy of the source data. Further, it is of value to recall that the theoretical bounds of LPIs are not fully understood, and that proportion (composition) and autocorrelation (configuration) have strong, and sometimes interactive and nonlinear, influences on LPI values.

The combination of spatially comprehensive and accurate disturbance information with pattern analysis allows for the provision of insights on the nature of the change to the landscape (and implications). For example, there is wide interest in determining the character of the relationships between natural and anthropogenic disturbances; questions arise, such as "Can natural disturbance type (fire) be emulated with forest management activities (e.g., harvest techniques)?" (e.g., Perrera et al. 2004). Stand-replacing harvest, road building, and fire disturbances are among the primary factors determining landscape pattern over the forests of North America. It is known that, prior to European settlement and significant timber-harvesting activity, fire was the principal disturbance shaping landscape pattern (Hudak et al., Chapter 8). The disturbance patterns currently evidenced on the landscape are largely at the patch size of typical harvest units, whereas historically, common disturbance patch sizes were quite different. Hudak et al. exemplify the remote sensing and GIS approach to test whether different disturbance regimes produce similar landscapes; in their own words, they set out to "demonstrate consistent and objective use of remote sensing and geographical information system (GIS) tools to characterize and compare the patch characteristics of stand-replacing harvest and fire disturbance processes in a coniferous forest landscape where both disturbances were known to have recently occurred." The approach was based on remotely sensed data with GIS support to map both harvest and fire disturbances. The authors employed change detection approaches that enabled the differentiation between forest change and condition over time. The results of their case study illustrate some interesting trends and differences regarding the LPI representing the harvest- and fire-disturbed areas. For example, they found that clearcutting results in forest patterns characterized by smaller patch sizes, smaller patch perimeter lengths, greater distances between patches, more edge habitat, and less interior habitat when compared to patterns created by natural processes such as fire, insect outbreak, avalanches, and blow-downs. Some management implications were made clearly evident in the patterns that emerged; for example, there were differences in patterns associated with own-ership of the burned areas. While some trends were evident, the authors were also careful to caution users to be diligent and transparent in method selection and application. Presentation of the pertinent issues and information needs, followed by a case study, aid in the understanding of how remote sensing and GIS can combine with pattern analysis approaches to produce otherwise unavailable information and allow for landscape-level inferences and comparisons and, over time, enable improved forest management. Clearly, improved understanding of the relationships between harvesting activities, ownership, forest structure, and resultant burn condi-tions is of high value to forest and land managers.

PLANNING FOR FUTURE DEVELOPMENTS

The future of pattern analysis appears tied to that of remote sensing; the satellite systems available will be capitalized on by the landscape ecology community to inform and develop a wide range of interests. This linkage is encouraging for forest managers, scientists, and landscape ecologists alike as the remote sensing community is poised to continue sensor developments (Stoney, 2004) and to create novel infor-mation generation methods. Gergel (Chapter 7) presents some notions regarding challenges and opportunities when considering co-developments of remote sensing and pattern analysis, including accuracy assessment of landscape metrics, pattern analysis using high spatial resolution imagery, and linear landscape metrics. Of these, pattern analysis using high spatial resolution imagery has the greatest potential for impact through increased data availability and lower acquisition costs. As mentioned by Gergel, many of the assumptions regarding the interpretation of LPI are based on the processing of Landsat imagery. Landsat pixels are composed of many ele-ments (i.e., trees, shadows, shrubs in a forested environment), whereas, high spatial resolution data, with pixel size 1×1 m or less, may be composed of a single, or few, elements. As a result, the relationships between pixels (configuration) will be changed, with the nature of the actual neighboring elements changed as well, poten-tially from stand components to tree components. Depending on the high spatial resolution sensor under consideration, the spectral resolution must also be consid-ered. The commercial systems currently available (Ikonos, QuickBird) collect pan-chromatic data with the high spatial resolution (1 m or less) channels onboard. Many of the image classification or change detection methods described in this book, and elsewhere, appear to operate best on multispectral channels, and these are (at present) typically associated with a lower spatial resolution (i.e., 4-m Ikonos, 2.4-m Quick-Bird). Further, the paradigm of many pixels per element described for the high spatial resolution situation does not necessarily hold for pixels of this (2.4- to 4-m) spatial resolution. The ability to robustly classify and detect change with high spatial resolution imagery is currently a major research focus in the remote sensing community (Wulder et al., 2004).

General advances in the ability to detect, map, and monitor disturbances, with the intention to process these disturbances subsequently with pattern analysis techniques to gain new insights into landscape processes, will be continually enhanced by ongoing increases in computing power, inexpensive disk storage, and advances to geospatial data and tools. Encouraging these trends is that satellite data are increasingly available to characterize large areas over long time periods with medium spatial resolution imagery, appropriate for landscape-level analyses (Franklin and Wulder, 2002).

LPIs typically consider only the horizontal distribution of structure. Although only mentioned briefly in this book, LIDAR (light detection and ranging) data are considered poised to lead to some breakthroughs in the ability to make ecological inferences through pattern analysis. The vertical structural component of data collected with LIDAR systems will enable patterns to be compared within locations (forest strata) and between locations. Measures of forest structural complexity, providing vertical information (LIDAR-based strata-level complexity) may enable development of vertical pattern stacks. Using the parlance of the GIS community, "voxels" of pattern may be developed, providing each location with additional complexity information that can be linked back to ecological conditions.

The capacity of GIS and database applications to store and manipulate increasingly large data structures opens opportunities for altering the relationship between remotely sensed and GIS data. Rogan and Miller (Chapter 6) indicated that the relationship was previously that of data producer (remote sensing) and data user (GIS). Rather than attempting to use remote sensing to produce and keep classifications current, a regular tessellation could be created (e.g., 25 m) by which land cover is stored and maintained. Changes to such a database could be integrated based on remote sensing of disturbance and through modeled growth. Rather than utilizing only remotely sensed depictions of landscape conditions, modeled conditions (based on spatial data, including remotely sensed data) capturing a wide range of detailed structural attributes can be perturbed with remote sensing depictions of change. The locations of change will be noted in the GIS, and the forest information will be initialized and "digitally grown" from that spatial location and point in time into the future. Monitoring in such a way will enable the reporting on forest processes in a spatially explicit manner. Development of such databases will allow for the observation of patterns that are currently present and those that are envisioned based on growth or differing change scenarios. Forest and land managers will have the ability to make decisions based on knowledge of current conditions and interactions as well as to "see" the impact of those management decisions (i.e., harvesting). Management decisions could be made based on particular landscape ecological conditions, as captured with LPI, that are deemed desirable in the future, for example, thresholds for fragmentation levels associated with roads and pipeline developments in a given forest management area. Hypothesis testing of differing disturbance scenarios would be enabled. Further, coupling of visualization tools with landscape conditions that represent the desired ecological conditions could be viewed and considered.

TOWARD FOREST SUSTAINABILITY

Remote sensing and GIS approaches are increasingly recognized, in combination with field data and modeling methods, as a realistic means to monitor landscape change over large areas with sufficient spatial detail to allow comparison of resultant patterns of different management or natural disturbance regimes. The approach, in general, consists of three broad but critically important steps: (a) understand the process of interest; (b) clarify the information needs required to make appropriate management decisions regarding the processes and patterns that result; and (c) implement appropriate remote sensing and GIS data selection and analytical methods to obtain optimal desired or specified output products.

Forest ecosystems are complex. Management of forest ecosystems is an enormously challenging undertaking, particularly with sometimes unclear limits to knowledge, yet few would argue that environmental management be conducted with less information. The impact of human management decisions on the forest landscape creates further complexities through interactions. As a result, the mapping and monitoring of disturbances must be accurate, and the LPIs generated through pattern analysis must be predictable, with known properties measured on the landscape. We articulate in this book the growing convergence of maturing disciplines — remote sensing, GIS, landscape ecology, forest management — that together generate a framework that enables disturbance mapping and subsequent pattern analysis of sufficient quality and consistency that a wide range of ecological and forest stewardship insights are made possible. These insights — for example, the comparison of landscape structure before or after one dominant disturbance regime (e.g., fire) has been replaced with another (e.g., harvesting) — result in an improved understanding of the linkages between existing patterns and processes operating on the landscape and changes to these patterns and processes over time. An increase in the ability and desire of forest managers to consider novel approaches has also occurred as a result of these new approaches to sometimes familiar yet seemingly intractable management problems. As indicated in this book, we believe many advances in remote sensing, GIS, and landscape ecology have been made, but clearly this is simply the beginning of a concentrated effort to bring coherence to previously isolated and diverse activities into a new, synergistic, and integrated approach. Ultimately, our hope is that by improving our collective ability to consider fully all of the competing factors that have an impact on our forests, whether those forests be managed principally for conservation or timber or for that matter any particular management objective, more insightful and inclusive management practices will emerge that will contribute toward long-term forest sustainability.

REFERENCES

Forman, R.T.T. (1995). *Land Mosaics, the Ecology of Landscapes and Regions*. Cambridge University Press, Cambridge, U.K. 632p.

Franklin, S.E. and Wulder, M.A. (2002). Remote sensing methods in medium spatial resolution satellite data land cover classification of large areas. *Progress in Physical Geography, 26*, 173–205.

Perrera, A.H., Buse, L.J., and Weber, M.G. (Eds.). (2004). *Emulating Natural Forest Landscape Disturbances: Concepts and Applications*. Columbia University Press, New York. 315p.

Stoney, W.E. (2004). *ASPRS Guide to Land Imaging Satellites*. Mitretek Systems. Retrieved April 22, 2005, from http://www.asprs.org/news/satellites/satellites.html.

Wulder, M.A., Hall, R.J., Coops, N.C., and Franklin, S.E. (2004). High spatial resolution remotely sensed data for ecosystem characterization. *Bioscience, 54*, 511–521.

Index